Crash Course in PC and Microcontroller Technology

Louis E. Frenzel, Jr.

Newnes

Boston Oxford Auckland Johannesburg Melbourne New Delhi

Newnes is an imprint of Butterworth-Heinemann.

Copyright © 1999 by Butterworth-Heinemann

A member of the Reed Elsevier group

 Recognizing the importance of preserving what has been written, Butterworth-Heinemann prints its books on acid-free paper whenever possible.

 Butterworth-Heinemann supports the efforts of American Forests and the Global ReLeaf program in its campaign for the betterment of trees, forests, and our environment.

Library of Congress Cataloging-in-Publication Data
Frenzel, Louis E.
Crash course in PC and microcontroller technology / Louis E.
    Frenzel.
        p.   cm.
    Includes index.
    ISBN 0-7506-9708-3 (pbk. : alk. paper)
    1. Microcomputers. 2. Programmable controllers. I. Title.
QA76.5.F744 1999
004.16—dc21   98-45597
                                                                CIP

British Library Cataloguing-in-Publication Data
A catalogue record for this book is available from the British Library.

The publisher offers special discounts on bulk orders of this book.
For information, please contact:
Manager of Special Sales
Butterworth-Heinemann
225 Wildwood Avenue
Woburn, MA 01801-2041
Tel: 781-904-2500
Fax: 781-904-2620

For information on all Butterworth-Heinemann publications
available, contact our World Wide Web home page at:
http://www.bh.com

10 9 8 7 6 5 4 3 2 1

Printed in the United States of America

Contents

Preface

If you want to invest in and read only one book on microcomputers this year, this book should be it. There is no faster way to bring yourself up to speed on personal computer and embedded controller fundamentals so that you can speak intelligently about them and understand the constant changes taking place in the electronics and computer fields.

So, how many computers have you used today? Maybe you used your computer at work, or perhaps you surfed the Net on your home PC. But you also used at least a half dozen other computers during the day without even realizing it. These "hidden" computers are what we call embedded microcontrollers, because they are built into just about every piece of electronic equipment that exists. Even though they are relatively simple single-chip computers, they are genuine digital computers with CPU, memory, I/O, and peripheral equipment. They are used to monitor and control virtually every electronic function imaginable.

If you have used any of these devices, you have used a computer: cell phone, pager, microwave oven, TV set, VCR, CD player, camcorder, burglar alarm, or dishwasher. Your car has several microcontrollers in it, typically one each for the ignition, fuel injection, the dashboard, and the anti-lock brakes. What's more, there are micros in the radio, tape player, CD player, security alarm, and the keyless remote entry (if you have one). Embedded controllers are everywhere. In fact, it's hard to name a piece of electronic equipment that does not use one.

Further, special integrated circuit microcomputers, called digital signal processors (DSP), are rapidly replacing more conventional analog circuits in many applications. DSP chips are showing up in cell phones, TV sets, CD players, modems, and PC sound boards where they replace traditionally analog circuits such as filters, modulators and demodulators, and equalizers. Not only do they perform better than the analog circuits they replace, DSP chips can do things that analog circuits could never do; a good example is spectrum analysis.

If you work in the electronic or computer fields, you must not only know how personal computers function, but you must know how the microcontrollers work as well. Such knowledge will satisfy your curiosity and give you a solid technical foundation on which to base your understanding of more advanced computer topics.

Content

The book begins with binary numbers, data, and codes, which are the language of computers. It then discusses digital computer hardware, including organization, architecture, and operation. Simple microprocessors are covered first, followed by memory (DRAM, SRAM, PROM, EEPROM, and flash) and then input/output circuits. Typical computers-on-a-chip or embedded microcontrollers are then discussed, including details on the most popular units such as the Motorola 68HC11, Intel 8051, Microchip Technology PIC16C56, and a Texas Instrument DSP chip.

PC circuits are also discussed. You will learn about the organization of the typical personal computer and about the Intel Pentium that is at the heart of over 90 percent of PCs. You will learn about buses such as the ISA, EISA, and PCI buses; I/O interfaces like RS-232, Centronics parallel, and SCSI; as well as the two new contenders USB and IEEE1394. PC peripherals such as printers, monitors, mouse, keyboard, disk drives, and others are also analyzed.

Any book on microcomputers would not be complete without coverage of the other half of any computer system, the software. The different types of software are described, such as operating systems, languages, and applications programs. You will discover how operating systems work, such as Windows 98 and UNIX. The fundamentals of languages such as BASIC, Fortran, Pascal, C++, and Java are covered. You will also learn programming fundamentals, including how to solve problems with algorithms, structured programming, and object-oriented programming (OOP). You will learn how to program a microcontroller in machine and assembly language. A chapter on the BASIC language is included to further hone your programming knowledge.

How to Use This Book

This book is a complete self-study course in microcomputer fundamentals. Each unit begins with a list of learning objectives that state what you will know and be able to do when you complete the unit. The topic to be learned is then presented in a unique programmed instruction format: The material is divided into frames, each of which covers a specific part of the subject. At the end of each frame there is a problem to solve or a question to answer to check your learning and understanding.

This book has been written and edited so that only the information you need is presented. The text has been boiled down to just the essential facts to keep the reading short, fast, and easily understandable.

In addition, this book also comes with a floppy disk that will work in any PC. The floppy disk contains interactive self-test review questions for each unit. These exams will help you to review and remember the key points in each chapter.

This book is the third in my Crash Course series. If you would like a complete course in electronics, be sure to look at my other two books, *Crash Course in Electronics Technology* and *Crash Course in Digital Technology*. Although you can benefit by reading this book by itself, you may want to get the whole story by including these other two books, which are an excellent prerequisite to learning about microcomputers.

Enjoy.

Lou Frenzel
Austin, Texas
1999

Microcomputer Fundamentals

Concepts and Definitions

1   Whenever you hear or see the word *microcomputer,* you probably think of a personal computer. A personal computer is, of course, one type of microcomputer. But you may not be aware of other forms of microcomputers. For example, do you know that almost every piece of electronic equipment you own or use has a microcomputer in it? This type of microcomputer is called an *embedded controller.* It is usually a single integrated circuit that performs all of the basic functions of a computer but is dedicated to a specific task. Such *micros* (*micro* is short for *microcomputer*) are in TV sets, stereo receivers, microwave ovens, CD players, VCRs, fax machines, cellular telephones, blenders, copiers, and many other types of electronic devices. You will find them in your car (several of them), your pager, your bathroom scale, gasoline pumps, and dozens

(continued next page)

of other things you use every day. Microcomputers do indeed make your day—you cannot live without them.

This book is about the various types of microcomputers and how they work. As a person interested in electronics, you need to know about microcomputers since they impact every aspect of electronic equipment. This book will tell you the nitty-gritty details about how micros are organized and programmed to do what they do. When you finish this book you will have a whole new appreciation for micros and how they perform their magic.

A common short term for a microcomputer is:

    a. microprocessor
    b. personal computer
    c. micro
    d. control unit

---

2   (*c.* micro)   A *microcomputer* is a small digital computer that can take several different forms. It can be a single large-scale integrated (LSI) circuit, or it can be a module made up of many integrated circuits on a printed-circuit board. The term microcomputer can also refer to a complete general-purpose digital computer system.

The two most common types of microcomputers are the *embedded microcontroller* and the *personal computer.* Some special types of microcomputers are handheld programmable electronic calculators and programmable logic controllers (PLCs) used for industrial control.

Regardless of its exact form or application, a microcomputer is still a(n)

    a. integrated circuit
    b. calculator
    c. digital computer
    d. controller

---

3   (*c.* digital computer)   Most microcomputers are made with a microprocessor. A *microprocessor* is a large-scale integrated circuit that contains most of the digital logic circuitry usually associated with a digital computer. This logic circuitry is referred to as the *central processing unit (CPU).* A microprocessor is a single-chip CPU. Digital computers that incorporate a microprocessor are said to be microprocessor-based. The key component in every microcomputer is the microprocessor.

A microprocessor is a single IC:

    a. central processing unit
    b. microcomputer
    c. memory
    d. digital computer

4   (*a.* central processing unit)   We will give you more precise definitions of the terms microprocessor and microcomputer later, but first let us learn more about a digital computer.

A *digital computer* is an assembly of electronic circuits that processes data. More specifically, a digital computer is an electronic system made up of digital logic circuits such as gates and flip-flops that are used to process data. The simple diagram in Figure 1-1 illustrates this idea. A digital computer is sometimes referred to as a data processor, or simply processor.

The main function of a microcomputer is to:

    a. replace discrete logic chips
    b. process data
    c. solve tough math problems
    d. manipulate text

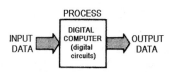

Fig. 1-1. **Simplified concept of a digital computer.**

---

5   (*b.* process data)   *Data* refers to any type of information such as numbers, letters, or special symbols. It may also refer to names, addresses, part numbers, words, sentences, and even paragraphs. Some examples of data are:

Types of Data

$976.38	(bank balance, pay check, etc.)
456-63-2975	(social security number)
233 Wall Street	(address)
314-726-4509	(telephone number)
49085	(zip code)
Your bill is one month overdue.	(message)
CA12	(PIN number for ATM access)
xyz@srf.org	(Internet Web address)

Most data tends to be numerical, but it can take practically any form. Data is any alphanumeric information. The terms *data* and *information* are used interchangeably.

The input to and output from a digital computer is called:

    a. numbers
    b. letters
    c. knowledge
    d. information

Ways Data Is Processed

6 (*d.* information)   *Processing* refers to the way the data is manipulated or handled. Types of processing include:

- Arithmetic
- Sorting
- Merging
- Logic
- Correcting
- Translating
- Editing
- Counting
- Compiling

Any action taken on the data is called processing. Processing normally implies that the data is changed in some way or is used to create new data.

Which of the following is *not* considered to be a type of processing?

  a. transmitting data from one place to another
  b. spell checking and correction
  c. calculating a percentage
  d. changing the order of words by alphabetizing

---

7 (*a.* transmitting data from one place to another) Data that is not processed can be dealt with in other ways. Four common ways are:

- storing
- retrieving
- input
- output

The data is not changed by any of these techniques.

Storing data means putting it in a safe place where it can be accessed later. *Retrieving,* of course, is the opposite of storing, or going to get the data for reuse.

Input means taking data into the computer to be stored or processed. *Output* means sending the data from the computer to some external device. Input and output are ways to transmit data from one place to another.

Saving data for later use is called:

  a. retrieval
  b. output
  c. storage
  d. input

8 (*c.* storage)   A good example of a data processing application with a microcomputer is common business transactions. Microcomputers can handle accounts receivable and payable, keep a general ledger, manage an inventory, print paychecks, update a mailing list, generate sales forecasts, and maintain employee records, among other things. In all of these cases, data must be entered via a keyboard, processing must take place, and appropriate output must be generated, usually with a printer. Personal computers are widely used for this kind of processing.

What must happen prior to processing in the business application just described?

> *a.* storage
> *b.* data input
> *c.* calculating
> *d.* output

---

9 (*b.* data input)   Although the major function of a digital computer is to process data, this characteristic is not unique to digital computers. For example, a handheld or desktop electronic calculator can also process data. Numbers (data) are entered by the keyboard and are used in some type of arithmetic operation (the processing), and an output result (data) is displayed on an electronic readout or is printed.

For this reason, can an electronic calculator be called a digital computer?

> *a.* Yes
> *b.* No

---

10 (*a.* Yes)   While an ordinary calculator is a computer according to our earlier definition, it is not a true computer because it does not have two key characteristics:

1. Automatic operation
2. Decision-making capability

Digital computers carry out their processing automatically without operator intervention once they have been given the input data. A real computer is self-regulating or self-controlling. Once you tell the computer to start processing, it will run to completion or continuously perform its function until you turn it off.

Automatic, as it applied to computers, means that the computer:

> *a.* is self-starting
> *b.* requires no power
> *c.* does whatever it wants
> *d.* needs no human assistance to keep it going

11   (*d.* needs no human assistance to keep it going) Another feature of a true digital computer is decision-making capability. During processing, the computer can make decisions and alter its sequence of operation. In other words, the computer can "make up its mind" based on the state of the data or outside conditions.

For example, the computer can tell if a number is less than, greater than, or equal to another. It can choose among alternative courses of actions or say yes or no, true or false, if given enough input facts.

An ordinary desktop calculator can make decisions.

     a. True
     b. False

---

12   (*b.* False)   Another major application of computers is control. Computers can be used to actuate relays and solenoids or turn lights and motors off and on. See Figure 1-2.

When a computer is used to control external devices, it is actually being employed as a sophisticated timer or sequencer. Many devices can be controlled at the same time by a single computer. Microcomputers are widely used in this type of application and are referred to as *microcontrollers.*

In control applications, the computer actually determines when external devices are turned on or off. The computer serves as an electronic clock to time various operations. One example is the embedded single-chip microcontroller inside a microwave oven or a washing machine. Another example is a programmable logic controller (PLC), a special type of microcomputer used in industrial applications, that is commonly used to sense, sequence, and time operations in a factory.

Microcontrollers and PLCs are used to:

     a. time operations
     b. actuate external devices
     c. initiate external operations
     d. all of the above

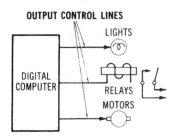

Fig. 1-2. A digital computer can control external devices.

---

13   (*d.* all of the above)   The most important thing to remember is that the control is automatic. The computer knows when to perform the various operations that it has been assigned because it has been programmed to do so.

In order to implement some control applications, the computer must also perform a monitoring function. The computer "looks at" the process or devices being controlled to see what is happening. For example, the computer can monitor switch closures to determine physical state or position, pressure, and many other parameters. Transducers (sensors) convert the monitored physical charac-

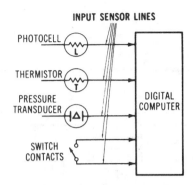

<figure>INPUT SENSOR LINES</figure>

Fig. 1-3. A digital computer can monitor external inputs.

teristics, such as temperature or light level, into electrical signals that the computer can understand and respond to. See Figure 1-3. In most microcontrollers or PLCs, the controlled output change takes place only if a specific input condition is sensed or not sensed. For example, if a temperature sensor indicates that the temperature has risen to a specific level, the computer will turn on a cooling fan.

When a computer is used for control, its outputs may occur as the result of input condition changes.

    a. always
    b. never
    c. sometimes

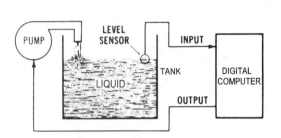

Fig. 1-4. A computer is used to control liquid level in a tank.

14   (*c.* sometimes)   Computers sometimes change the control function in response to one or more of the inputs it is monitoring. For example, if the computer senses that the liquid in a tank exceeds a given level, it can automatically turn off the pump that is filling the tank. If the liquid level goes below a predetermined level, the computer detects this and automatically starts the pump. See Figure 1-4. The key point here is that the computer makes its own decisions based on the input data it receives. The result is full automation of some process or device.

The computer's ability to monitor and control an operation may be used to automate simple processes or to operate entire factories.

A microcontroller is used to control the speed of a 1/4-horsepower (hp) motor. The computer output is a signal that is sent to the motor to set its speed at a specific value. The input to the computer is:

    a. a count of shaft turns in degrees
    b. motor shaft rpm
    c. a measure of hp
    d. a measure of torque

---

15   (*b.* motor shaft rpm)   In summary, we can say that a digital computer is an electronic machine that:

    1. processes data
    2. makes decisions
    3. operates automatically
    4. performs monitoring and control functions

Computer Organization and Operation

16 All digital computers are made up of four basic sections: the *memory,* the *control unit,* the *arithmetic-logic unit,* and the *input-output (I/O) unit.* A general block diagram of a digital computer showing these four sections is given in Figure 1-5.

These four sections communicate with one another over multiple parallel electrical conductor data paths called a *bus,* as shown, to process the data or perform some control function.

Note that the control and arithmetic-logic unit (ALU) are shown together in a common structure. This is the central processing unit (CPU). The CPU is, of course, usually a microprocessor.

Communications between the main sections of a computer take place over electrical connections called a(n):

    a. cable
    b. data route
    c. bus
    d. wire

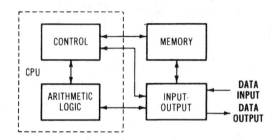

Fig. 1-5. General block diagram of a digital computer.

17 (*c.* bus) The *memory* is that part of the computer where data and programs are stored. The memory in any computer may contain thousands or even millions of locations used for storing numbers, words, or other forms of information.

Two primary types of information are stored in computer memory. The first type is the data to be processed. These are the numbers, letters, and other forms of data to be manipulated.

The other type of data stored in memory is an *instruction.* Instructions are special numbers or codes that tell the computer what to do. Instructions specify the ways in which the data is to be processed. There are instructions that cause arithmetic operations to take place or data to be transferred from one place to another.

The code that designates how data is processed is called:

    a. an instruction
    b. data
    c. a program
    d. software

18 (*a.* an instruction) The instructions listed in a special sequence form a program. A *program* is a list of instructions that causes data to be processed in a unique way. A program is a step-by-step procedure that solves a problem, performs a control operation, or otherwise manipulates data according to some recipe. Programs are stored in memory along with the data that they process. The programs that a computer uses are called *software*.

A program is:

    a. a list of data values
    b. the instructions a computer can perform
    c. special codes to perform specific operations
    d. a sequence of instructions that does something useful

19

(*d.* a sequence of instructions that does something useful)   Data and instructions can be stored in or retrieved from memory during processing. When data is stored, we say that it is written into memory. When data is retrieved, we say that it is read from memory. A typical write operation transfers a program and data from a disk drive and stores it into memory. A common read operation accesses data in memory to be transferred to a printer or the video screen.

Loading a program into memory calls for a:

    a. read operation
    b. write operation

20

(*b.* write operation)   Now let's look at the operation of the CPU. The *control unit* is that portion of the digital computer responsible for the automatic operation. The control unit sequentially examines the instructions in a program and issues signals to the other sections of the computer that carry out the designated operations.

Each instruction is fetched (read) from memory by the control unit, interpreted, and then executed one at a time until the program is completed. This is called the *fetch-execute cycle,* which is repeated on each instruction until the program runs to completion. The execution of each instruction may call for accessing one or more data words in the memory or storing a data word in memory.

The control section's repeated fetch-execute cycle leads to:

    a. a calculated solution
    b. a write operation
    c. automatic program operation
    d. an I/O operation

21

(*c.* automatic program operation)   The *arithmetic-logic unit (ALU)* is the section of the computer that carries out many of the functions that are specified by the instructions. In other words, the ALU actually processes the data. Specifically, the ALU carries out two main types

(continued next page)

of processing: arithmetic operations (such as addition, subtraction, multiplication, and division) or logic operations (such as AND, OR, complement, or exclusive OR). For example, if an "add" instruction is stored in memory, the control section will fetch it, interpret it, and send signals to the ALU that cause two numbers to be added.

The ALU also performs data movement operations. It can move data or instruction words from one place to another inside the CPU or it can carry out memory read/write or input/output operations.

Which section of the computer actually does the "processing"?

    a. ALU
    b. I/O
    c. control
    d. memory

22

(*a.* ALU)   A key part of the CPU associated with the ALU is the registers. *Registers* are digital circuits made up of flip-flops that store a binary word or number. Most ALUs have two or more registers. High-powered CPUs may have a group of 16 or more registers referred to as *general purpose registers.* The registers are used to temporarily store the data being processed by the ALU and the results of the computations. Other registers in the CPU store a number called an *address* where the data or instructions are stored, store the instruction being executed, or act as a stop-off place for data into or out of the CPU.

A register does which of the following?

    a. performs logic
    b. performs arithmetic
    c. stores instructions or data
    d. automates I/O operations

23

(*c.* stores instructions or data)   The control section, ALU, and the registers in all digital computers are very closely related. They operate together and are always considered as a single unit. As indicated earlier, the combination of the control and ALU sections is called the central processing unit (CPU). See Figure 1-5. In addition, microprocessors are single-chip CPUs. Besides being called CPUs, microprocessors are sometimes called *MPUs* or *microprocessing units.* You will also see the expression μP used to refer to a microprocessor. The "μ" is the Greek letter mu, which means micro; the P means processor.

Which of the following is *not* a common name for a microprocessor?

    a. CPU
    b. microcomputer
    c. μP
    d. MPU

24 (*b.* microcomputer)   A microcomputer consists of a microprocessor (the CPU) plus the memory and I/O circuits.

The *input/output (I/O) unit* of the computer is the set of logic circuits that permits the CPU and memory to communicate with the outside world. All data transfers into and out of the computer pass through the I/O unit. The I/O unit acts as the interface between the computer and any external peripheral device.

The *external peripheral devices,* or *peripherals,* connected to the I/O unit are electronic or electromechanical units that are used for data entry or data display. Data is most commonly entered into the computer through an input keyboard. A mouse is another popular input device.

The output data is usually displayed on a video monitor cathode-ray tube (CRT). "Hard copy" on paper is created by a printer. There are a wide variety of other external input/output devices such as scanners, voice synthesizers, bar code readers, and plotters.

The two most common peripherals on a computer are:

    a. keyboard, video monitor
    b. mouse, printer
    c. scanner, plotter
    d. keyboard, printer

25 (*a.* keyboard, video monitor)   Another type of peripheral device is the *auxiliary memory.* These devices are used for mass data storage. Typical auxiliary memory media are magnetic tapes, floppy disks, hard disks, CD-ROMs, and digital versatile disks (DVDs).

Most computers have a built-in floppy disk (diskette) drive and a *hard drive,* also called a *fixed disk.* Magnetic tape units are used primarily for backing up (protecting) the contents of the hard drive. CD-ROM drives store massive amounts of data and programs and are widely used for multimedia applications involving computer text, graphics, video, and audio. The DVD is a disk like a CD-ROM but it stores much more data and can also store huge amounts of video and audio information.

Disk and tape drives are most accurately referred to as:

    a. I/O devices
    b. peripherals
    c. extra memory
    d. auxiliary mass storage

26 (*d.* auxiliary mass storage)   The digital computer combined with its peripherals forms a complete

(continued next page)

computer system for data processing or control applications.

In an embedded microcontrollers, the peripheral devices include:

Inputs

- Keyboard
- Barcode scanner
- Magnetic tape stripe reader (as on a credit card)
- Sensors to measure temperature, pressure, light, position, and so forth

Outputs

- Relays
- Lights
- Solenoids
- Motors
- 7-segment LED display readouts
- Cathode-ray tubes

Now answer the Self-Test Review Questions on the diskette before going on to the next unit.

Binary Data

LEARNING OBJECTIVES

When you complete this unit, you will be able to:

1. Convert binary numbers to decimal numbers.

2. Convert decimal numbers to binary numbers.

3. Explain why binary data is preferred in computers over decimal data.

4. Define the terms *base, radix, most significant digit, least significant digit, word, weight, bit,* and *byte.*

5. Convert binary numbers into hex.

6. Convert hex numbers to binary.

7. Convert decimal numbers to BCD.

8. Convert BCD to decimal numbers.

9. Make data conversions to and from ASCII.

10. Add and subtract binary numbers.

11. Express negative numbers in binary.

12. Express fractional numbers in binary.

13. Define *floating-point numbers.*

14. Convert a floating-point binary number to decimal.

Number Systems

1   Numbers and codes are the primary language of all digital computers. The data that computers process is usually numerical in nature. Even alphabetic and symbolic

(continued next page)

data is expressed as numerical codes. Because of this, you must be familiar with the number systems that express the various ways that data is represented and manipulated in a computer.

Computers use a special number system to represent quantities and process them. Because this system uses only two symbols or digits to represent the quantities 0 and 1, it is called the *binary number system*. Binary numbers are more easily processed by a computer than other numbers. All digital computers use binary numbers and codes.

Binary means:

 a. decimal
 b. two
 c. code
 d. digital

---

2 (*b. two*) The types of numbers with which we are most familiar are called *decimal numbers*. The decimal number system is a method of communicating numerical information by using the symbols 0 through 9. These ten symbols or digits can be combined in a variety of ways to represent any quantity.

The distinguishing characteristic of any number system is its *base* or *radix*. The base is the number of characters or symbols used to represent quantities. The base of the decimal number system is ten.

What is the radix of a number system that uses the characters 0 through 5 to represent quantities?

 a. 2
 b. 5
 c. 6
 d. 10

---

3 (*c. 6*) The decimal number system is a positional or weighted system. That is, each digit position in a number carries a specific weight in determining the magnitude of that number. The position weights of a decimal number are units, tens, hundreds, thousands, and so on. This is illustrated in the example that follows. Consider the decimal number 5,692. Note that the most significant digit (MSD) and the least significant digit (LSD) are identified.

	Position Weight	Digit Value	
MSD	Thousands	5	× 1,000 = 5,000
	Hundreds	6	× 100 = 600
	Tens	9	× 10 = 90
LSD	Units	2	× 1 = 2
			5,692

Note that the weight of each position is multiplied by the corresponding digit to obtain the value of that position.

The values of each position are then summed to obtain the original number. In any number, because the digit furthest to the right adds the least value, it is called the *least significant digit (LSD)*. Because the digit furthest to the left adds the most value, it is called the *most significant digit (MSD)*.

In the number 2,310,798, the MSD has the weight of:

    a. 1,000
    b. 10,000
    c. 100,000
    d. 1,000,000

And the least significant digit is:

    e. 2
    f. 8

---

4   (*d.* 1,000,000; *e.* 8)   We can also express the position weights in powers of ten. Each position weight is the base of the number system raised to a specific power.

Units	$1 = 10^0$
Tens	$10 = 10^1$
Hundreds	$100 = 10^2$
Thousands	$1,000 = 10^3$
Tens of thousands	$10,000 = 10^4$
Hundreds of thousands	$100,000 = 10^5$
Millions	$1,000,000 = 10^6$

The powers of 10, as you may recall, are simply a shorthand system for expressing large quantities. Using powers of ten, the number 5,692 is written as:

$$(5 \times 10^3) + (6 \times 10^2) + (9 \times 10^1) + (2 \times 10^0) =$$
$$(5 \times 1000) + (6 \times 100) + (9 \times 10) + (2 \times 1) =$$
$$5000 + 600 + 90 + 2 = 5692$$

The number represented by the following expression is:
$$(1 \times 10^4 + 0 \times 10^3 + 4 \times 10^2 + 3 \times 10^1 + 7 \times 10^0)$$

    a. 14,037
    b. 10,437
    c. 73,401
    d. 104,037

---

5   (*b.* 10,437)   While humans use the decimal number system, computers do not. That is, the decimal number system is not used inside the computer. Computers do receive decimal numbers as inputs and produce decimal-number outputs to accommodate the human operator. But the computer does not process decimal numbers. Instead, digital computers use the binary number system.

The *binary number system* is a set of rules and procedures for representing and processing numerical quantities

(continued next page)

in digital computers. In fact, binary numbers are the basic language of all digital equipment. Since the base of the binary number system is two, only two symbols (0 and 1) are used to represent any quantity. The symbols 0 and 1 are called *binary digits* or *bits*. For example, the 6-bit number 101101 represents the decimal value 45.

Computers store and process decimal values.

    a. True
    b. False

---

6   (*b.* False)   The reason for using binary numbers in digital equipment is the ease with which they can be implemented. The electronic components and circuits used to represent and process binary data must be capable of assuming two discrete states to represent 0 and 1. Binary or two-state circuits are small, simple, fast, and economical. On the other hand, decimal circuits (those with ten states) are far more complex and costly. Examples of two-state components are switches and transistors. When a switch is closed or on, it can represent a binary 1. When the switch is open or off, it can represent a binary 0. A conducting transistor may represent a 1, while a cut-off transistor may represent a 0, or vice versa. The representation may also be voltage levels. For example, a binary 1 may be represented by +3 volts and a binary 0 by zero volts.

Which of the following is *not* a reason why binary representation in a computer is beneficial? Binary circuits are:

    a. large
    b. fast
    c. low cost
    d. simple

---

7   (*a.* large)   The binary system is similar to the decimal system in that the position of a digit in a number determines its *weight*. The position weights of a binary number are also powers of the number system base. In the binary system, each bit position carries a weight that is some power of 2. These are:

$2^0 = 1$ $2^1 = 2$ $2^2 = 4$ $2^3 = 8$
$2^4 = 16$ $2^5 = 32$ $2^6 = 64$ $2^7 = 128$

Study these binary position weights and determine how they are related to one another. Then answer the following question.

The next highest power of two that follows the highest one shown above is:

    a. 256
    b. 512
    c. 1,024
    d. 2,048

8   (a. 256)   The weight of each position is twice the value of the next lower position. This makes the position weights in a binary number easy to remember or determine. The position weights of an 8-bit binary number are illustrated as shown. Note that the most significant digit or bit (MSB) and the least significant bit (LSB) are identified.

Bit position	2^7	2^6	2^5	2^4	2^3	2^2	2^1	2^0
Weight	128	64	32	16	8	4	2	1
Example binary number	1	1	0	0	1	0	1	0
Significant bits	MSB							LSB

In a 12-bit binary number, the MSB will have a weight of:

    a. 512
    b. 1,024
    c. 2,048
    d. 4,096

9   (c. 2,048)   Go to the next frame.

Binary-Decimal Conversions

10   Now let's evaluate the decimal quantity associated with a given binary number. You can do this as you did with the decimal number earlier. You simply multiply each bit by its position weight and sum the values to get the decimal equivalent.

Position weights	32	16	8	4	2	1
Binary number	1	0	1	1	0	1

Since the positions with weights of 2 and 16 have 0s in them, multiplying will just produce zero, so these positions contribute nothing to the value of the number. The other positions have binary 1s in them, so we just add up the weights:

Decimal equivalent = 32 + 8 + 4 + 1 = 45

Using this technique, determine the decimal equivalent of the binary number 1001010.

(continued next page)

a. 41
b. 47
c. 74
d. 122

11

(c. 74)   The correct solution is given below.

Position weight	64	32	16	8	4	2	1
Binary number	1	0	0	1	0	1	0

0 x 1	=	0
1 x 2	=	2
0 x 4	=	0
1 x 8	=	8
0 x 16	=	0
0 x 32	=	0
+1 x 64	=	64
Total		74

You can see that those positions with a 0 bit have no effect on the value. Therefore, they can be ignored. To quickly determine the decimal equivalent of a binary number, you simply sum the weights of those positions containing a 1 bit. For example, in the number 11101, the weights of those positions with 1 bit from right to left are 1, 4, 8, and 16. This gives a decimal equivalent of 1 + 4 + 8 + 16 = 29.

What is the decimal equivalent of the number 101011?

a. 18
b. 43
c. 47
d. 53

12

(b. 43)   The weights of the positions with 1 bits are 1, 2, 8, and 32, for a sum of 43.

Switches are widely used to enter binary data into computers. A common example is a DIP switch. Each switch represents one bit of the binary number. The switches are set to either their binary 1 or binary 0 position to represent the desired number. Usually, if the switch is set to the up or on position, a binary 1 is represented. If the switch is down or off, a binary 0 is represented.

Figure 2-1 shows a group of slide switches set to represent a binary number. What is its decimal equivalent?

a. 87
b. 98
c. 163
d. 197

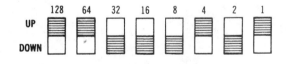

Fig. 2-1.   Binary data switches.

BINARY LIGHTS

ON (1)　　OFF (0)

Fig. 2-2. Binary lights.

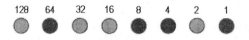

128　64　32　16　8　4　2　1

Fig. 2-3. Binary number display.

13 *(d. 197)* The 128, 64, 4, and 1 weight positions contain binary 1s. Therefore, the decimal equivalent is 128 + 64 + 4 + 1 = 197.

Indicator lights such as light-emitting diodes (LEDs) are often used to read or display binary data in a computer. An on light is a binary 1 and an off light is a binary 0. See Figure 2-2.

What decimal number is being represented by the display in Figure 2-3?

    a. 178
    b. 182
    c. 199
    d. 204

14 *(a. 178)* The on lights (binary 1) appear in the 128, 32,16, and 2 weight positions giving a total of 178. As you can see, converting binary numbers to decimal is easy. Simply sum the weights of the positions containing binary 1s.

You will also find it necessary to convert decimal numbers into their binary equivalents. This can be done by repeatedly dividing the decimal number by 2 (the base of the binary system) and noting the remainder. The remainder forms the binary equivalent.

This procedure is best illustrated by an example: Let's convert the decimal number 57 to its binary equivalent.

Quotient	Remainder
$57 \div 2 = 28$	1 (LSB)
$28 \div 2 = 14$	0
$14 \div 2 = 7$	0
$7 \div 2 = 3$	1
$3 \div 2 = 1$	1
$1 \div 2 = 0$	1 (MSB)

The number 57 is initially divided by 2. This gives a quotient of 28 with a remainder of 1. The quotient 28 is then divided by 2, giving 14 which, in turn, is divided by 2, and so on. This procedure is repeated until the quotient is zero. The remainders from each division form the binary number. Therefore, the binary equivalent of 57 is 111001.

Binary equivalent of 86 is:

    a. 0110101
    b. 1010110
    c. 1001011
    d. 1101000

15 (*b.* 1010110)   Your answer should look like this:

```
86 ÷ 2 = 43     0 (LSB)
43 ÷ 2 = 21     1
21 ÷ 2 = 10     1
10 ÷ 2 = 5      0
 5 ÷ 2 = 2      1
 2 ÷ 2 = 1      0
 1 ÷ 2 = 0      1 (MSB)
```

You can always check your work by reconverting the binary number to decimal using the procedure described earlier.

The binary equivalent of 16 is:

    a. 000100
    b. 001000
    c. 01000
    d. 10000

---

16 (*d.* 10000)   Another name for binary number is *binary word.* The term *word* is more general and can mean either numbers, letters, or special characters and codes. We say that the computer stores and processes binary data words.

A common specification of any computer or microcomputer is its word length. All microcomputers work with a fixed-length binary word. That is, the data words in the computer have a specific number of bits. Common binary word lengths in microcomputers are 8, 16, 32, and 64 bits. Most data storage, processing, manipulation, and transmission is carried out on word-length data.

In computer lingo, an 8-bit word is called a *byte.* A 4-bit word is sometimes referred to as a "nibble."

How many bytes are there in a 32-bit word?

    a. 1
    b. 3
    c. 4
    d. 8

---

17 (*c.* 4)   The number of bits in a binary word determines the maximum decimal value that can be represented by that word. This maximum value is determined with the simple formula:

$$M = 2^N - 1$$

M is the maximum decimal value while N is the number of bits in the word. For example, what is the largest decimal number that can be represented with 4 bits?

$$M = 2^N - 1$$
$$M = 2^4 - 1 \qquad (2^4 = 2 \times 2 \times 2 \times 2 = 16)$$
$$M = 16 - 1 = 15$$

With 4 bits, the maximum possible number is binary 1111 or 15.

The maximum decimal number that can be represented with one byte is:

    a. 63
    b. 127
    c. 255
    d. 512

---

18   (*c.* 255)   One byte has 8 bits, therefore:

$$M = 2^N - 1$$
$$M = 2^8 - 1$$
$$M = 256 - 1 = 255$$

An 8-bit word greatly restricts the range of numbers that can be accommodated in a computer. But this problem is usually overcome by using larger words or by using more than one word to represent a number. For example, two bytes can be used to form a single 16-bit word. The eight most significant bits are contained in one byte and the eight least significant bits in the other byte. See Figure 2-4. With 16 bits we can represent the maximum number $2^{16} - 1 = 65,535$. More words can be used if greater quantities must be represented. Larger microprocessors have word sizes of 32 and 64 bits.

How many 16-bit words are required to represent a 32-bit number?

    a. 1
    b. 2
    c. 3
    d. 4

1ST BYTE	1 0 1 0 1 1 1 0	MOST SIGNIFICANT BITS
2ND BYTE	0 1 0 0 0 1 0 1	LEAST SIGNIFICANT BITS

16 BIT BINARY WORD = 1 0 1 0 1 1 1 0 0 1 0 0 0 1 0 1

Fig. 2-4. Two 8-bit bytes form a single 16-bit word.

---

19   (*b.* 2)   There is one important point to note before leaving this subject. The formula $M = 2^N - 1$ determines the maximum decimal quantity (M) that can be represented with a binary word of N bits. This value is one less than the maximum number of values that can be represented. The maximum number of values that can be represented (Q) is determined by the formula $Q = 2^N$. Again N is the number of bits. For example, with 4 bits, Q is $2^4 = 16$. With 4 bits, 16 values can be represented. These 16 values are 0 through 15 where 15 is the maximum number ($2^4 - 1 = 15$). Note that zero is a valid value.

(continued next page)

With 6 bits, the total number of values that can be represented is:

    a. 63
    b. 64
    c. 127
    d. 128

---

20   (*b.* 64)   Table 2-1 gives the number of bits in a binary number and the maximum number of states that can be represented. Note also the designation, which is an abbreviation used to specify each quantity.

Table 2-1. **Common Binary Word Sizes and the Maximum Numbers of States that Can Be Represented**

Number of bits (N)	Maximum states (2^N)	Designation
8	256	256
12	4,096	4k
16	65,536	64k
20	1,048,576	1M
24	16,777,216	16M
32	4,294,967,296	4G

The abbreviation k means 1,024. Therefore, a quantity of 64k is 64 × 1,024 = 65,536. 1M means one mega (million) where M = 1,048,576. A value of 16M is 16 × 1,048,576 = 16,777,216. G means giga or one billion. Specifically, one G = 1,073,741,824. 4G = 4 × 1,073,741,824 = 4,294,967,296.

What is the exact quantity associated with the value 4M?

    a. 4,000,000
    b. 4,194,304
    c. 4,000,000,000
    d. 4,294,967,296

---

21   (*b.* 4,194,304)   Go to the next frame.

Hexadecimal Numbers

---

22   Because binary numbers are made up of only two symbols, they are rather difficult to remember. Long sequences of 0s and 1s are troublesome to work with. Even short binary numbers are tough to remember. Although it is possible to memorize a given 8-bit sequence like 10111010, it may be necessary to recall dozens of such numbers when working with microcomputers. Remembering 16, 32, or 64-bit numbers is next to impossible. Because of this, shorthand techniques have been developed to facilitate the use of binary numbers.

One of the techniques involves the use of the hexadecimal number system. Hexadecimal essentially means 6+10, or 16. The base of the hexadecimal system is 16.

The hexadecimal or hex system uses 16 symbols to represent quantities. These are the numbers 0 through 9 and the letters A through F.

The binary and hex systems are related in that the hex system base is a power of 2 ($2^4 = 16$). With 4 bits, 16 different numbers can be defined ($2^N = 2^4 = 16$). These 16 numbers are 0–9 and A–F. The binary and hex equivalents are given in Table 2-2.

Table 2-2. Hex and Binary Equivalents

Hex	Binary
0	0000
1	0001
2	0010
3	0011
4	0100
5	0101
6	0110
7	0111
8	1000
9	1001
10	1010
11	1011
12	1100
13	1101
14	1110
15	1111

The first 10 codes are the same as the decimal equivalents.

    a. True
    b. False

---

23   (*a.* True)   To use the hex system to simplify the representation of binary numbers, you divide the binary number into 4-bit groups starting with the LSB. Then replace each 4-bit group with its hex equivalent. For example, 10100101 becomes:

$$1010\,/\,0101$$
$$A\qquad 5$$

Note that the slash above does *not* indicate division as it usually does. Instead, it is simply used as a tool to visually divide a long binary number into four bit groups. Since 1010 = A and 0101 = 5, the binary number above is A5 in hex.

Here's another example:

(continued next page)

1000100110101101 becomes
1000/1001/1010/1101 or 89AD.

The hex equivalent of 0011100011110110 is:

    a. 04D9
    b. 6F83
    c. 2B17
    d. 38F6

24

(*d.* 38F6)   The letters within a number are somewhat strange, but you have to admit that a 4-digit hex number is easier to remember than a 16-bit binary number.

To convert hex numbers into binary, the substitution process is reversed. Each hex digit is replaced by its 4-bit binary equivalent. To convert 4C7E to binary, replace each hex digit with its 4-bit equivalent.

4	C	7	E
0100	1100	0111	1110

What is D80B in binary?

    a. 1101100000001011
    b. 1101000000011011
    c. 0101100111011010
    d. 0010011011100011

25

(*a.* 1101100000001011)   Computer designers and programmers usually work with three number systems: binary, hex, and decimal. Occasionally confusion develops because a given number can be interpreted in several ways. For example, just what is the value of 100? Is this a decimal, binary, or hex number? Of course, a different quantity would be represented in each number system. To eliminate this confusion, a subscript is sometimes added to the number to indicate which number system is being used. This subscript is the base of the number system. For example, 100_{10} means a decimal one hundred. The number 100_2 is the binary equivalent of the decimal number 4. The hex number 100_{16} is 100000000 in binary or a decimal 256.

Another method used to designate the base of a number is to place a letter suffix after the number. The letter H after a number means it is a hex number. Thus, 207FH is hex. B designates binary numbers, and a D designates decimal numbers.

A $ prefix is also sometimes used to designate a hex number. The value 207FH above written this way is $207F.

Give the decimal equivalents of the following numbers:

        11010010B = _____ D
           A5D3H = _____ D
            $1F02 = _____ D

26 *(See answers that follow.)*

$$11010010B = 210D$$
$$A5D3H = 42451D$$
$$\$1F02 = 7938D$$

In solving these problems you probably discovered that the best way to convert hex numbers to decimal is to convert them to binary first, then decimal.

Go to the next frame.

BCD and ASCII

27 The binary numbers we have been discussing are usually referred to as pure binary numbers or pure binary codes. But there are other types of binary codes. The pure binary code is the most widely used, but the binary coded decimal (BCD) system is nearly as popular.

The BCD system is essentially a cross between the binary and decimal systems. It was developed in an attempt to simplify the conversion processes between the two systems and to improve man–machine communications.

To represent a decimal number in the BCD system, each digit is replaced by its 4-bit binary equivalent. The BCD code is given in Table 2-3. Thus, the number 729_{10} in BCD is:

7	2	9
0111	0010	1001

When representing a number in BCD, spaces are used between each group of 4-bits. It is important to note that the 4-bit binary numbers 1010 through 1111 representing decimals 10 through 15 are invalid in BCD, therefore they are missing from Table 2-3.

Table 2-3. The BCD Code

Decimal	BCD
0	0000
1	0001
2	0010
3	0011
4	0100
5	0101
6	0110
7	0111
8	1000
9	1001

(continued next page)

Express 5031 in BCD.

    a. 1010 1111 1100 1110
    b. 0101 0000 0011 0001
    c. 0001 0011 0000 0101
    d. 0101 0011 0000 0001

28

(*b.* 0101 0000 0011 0001)   Note that each digit is replaced by its full 4-bit equivalent (including any leading zeros) and the groups are spaced to keep the digits separate.

To convert a BCD number to decimal, you simply substitute the decimal equivalent of each 4-bit group.

The BCD number 1001 0100 0110 in decimal is:

    a. 946
    b. 649
    c. 2,374
    d. 6,490

29

(*a.* 946)   You should memorize Table 2-3 to facilitate the conversions.

There is one important point you should note. It takes fewer bits to represent a number in pure binary code than in BCD. For example, 86 in binary is the 7-bit number 1010110. In BCD, 86 is the 8-bit number 1000 0110. The binary code is more efficient because it uses fewer bits. This can lead to a hardware savings in some applications. However, often this savings is traded off for the improved man–machine communications. Binary-coded decimal is widely used in digital equipment with decimal displays such as counters, TV and radio channel or frequency displays, digital voltmeters, and digital clocks.

Binary-coded decimal is a compromise code that is widely used with computers and microprocessors. Most microprocessors can store and process BCD data even though they are pure binary machines.

How many BCD digits can be contained with one byte?

    a. 1
    b. 2
    c. 3
    d. 4

30

(*b.* 2)   A byte is eight bits and each BCD character is four bits long.

A special form of BCD code is used in many computers and data-communications systems. It is a 7- or 8-bit code that is used to represent not only numbers, but also letters (both upper and lowercase), special symbols, and control functions. This code is called the *American Standard Code for Information Interchange* or *ASCII* (pronounced "ass key"). Most computers and microprocessors communicate with their peripheral equipment in the ASCII code. And virtually all on-line communications over the Internet are written in ASCII code.

The complete ASCII code is given in Table 2-4. There are both 7- and 8-bit versions, but the 7-bit version is more widely used. Leading zeros are used to complete the most significant hex digit. Referring to the table, you can see that the 7-bit ASCII code for the letter F is 46H or 01000110B.

Table 2-4. American Standard Code for Information Interchange

The rightmost column (C) is the ASCII character or control designation. The leftmost column (D) is the decimal equivalent. The next column is the hex (H) designation. For example, the ASCII character Y in hex is 59. This translates to a binary value of 01011001 and a decimal value of 89.

D	H	C	D	H	C	D	H	C	D	H	C
0	00	NUL	32	20	SP	64	40	@	96	60	'
1	01	SOH	33	21	!	65	41	A	97	61	a
2	02	STX	34	22	"	66	42	B	98	62	b
3	03	ETX	35	23	#	67	43	C	99	63	c
4	04	EOT	36	24	$	68	44	D	100	64	d
5	05	ENQ	37	25	%	69	45	E	101	65	e
6	06	ACK	38	26	&	70	46	F	102	66	f
7	07	BEL	39	27	'	71	47	G	103	67	g
8	08	BS	40	28	(	72	48	H	104	68	h
9	09	HT	41	29	)	73	49	I	105	69	i
10	0A	LF	42	2A	*	74	4A	J	106	6A	j
11	0B	VT	43	2B	+	75	4B	K	107	6B	k
12	0C	FF	44	2C	,	76	4C	L	108	6C	l
13	0D	CR	45	2D	-	77	4D	M	109	6D	m
14	0E	SO	46	2E	.	78	4E	N	110	6E	n
15	0F	SI	47	2F	/	89	4F	O	111	6F	o
16	10	DLE	48	30	0	80	50	P	112	70	p
17	11	DC1	49	31	1	81	51	Q	113	71	q
18	12	DC2	50	32	2	82	52	R	114	72	r
19	13	DC3	51	33	3	83	53	S	115	73	s
20	14	DC4	52	34	4	84	54	T	116	74	t
21	15	NAK	53	35	5	85	55	U	117	75	u
22	16	SYN	54	36	6	86	56	V	118	76	v
23	17	ETB	55	37	7	87	57	W	119	77	w
24	18	CAN	56	38	8	88	58	X	120	78	x
25	19	EM	57	39	9	89	59	Y	121	79	y
26	1A	SUB	58	3A	:	90	5A	Z	122	7A	z
27	1B	ESC	59	3B	;	91	5B	[	123	7B	{
28	1C	FS	60	3C	<	92	5C	\	124	7C		
29	1D	GS	61	3D	=	93	5D	]	125	7D	}
30	1E	RS	62	3E	>	94	5E	^	126	7E	~
31	1F	US	63	3F	?	95	5F	—	127	7F	DEL

What is the binary ASCII code for the decimal number 8?

    a. 00000111
    b. 00001000
    c. 00111000
    d. 01010110

31   (*c.* 00001000)   If you convert each of the ASCII codes for the digits 0 through 9 to binary, you will see that the last four bits of the code correspond to the BCD equivalent. Verify this yourself in Table 2-4.

What character is represented by the binary word 01001010?

      a. k
      b. J
      c. L
      d. $

---

32   (*b.* J)   You should have converted the binary number to hex, then looked up its equivalent in Table 2-4.

In Table 2-4, you will see that the ASCII code also contains some characters designated by two or three letters or by two letters and a number. Some examples are BS, ESC, and DC3. These are called *control codes.* These codes are interpreted by the hardware and will make things happen. For instance, in a printer BS will cause a back space and LF will cause a line feed. BEL causes a bell to ring. ACK means acknowledge, STX means start transmission, and so on.

Looking through Table 2-4, what ASCII character in hex might cause a delete operation?

      a. 00
      b. 10
      c. 44
      d. 7F

---

33   (*d.* 7F)   Another special BCD code used in computer-to-peripheral communications is *Extended Binary Coded Decimal Interchange Code* or *EBCDIC* (pronounced "ebb see dick"). EBCDIC is similar to ASCII but it is an 8-bit code. Like ASCII, it is used to represent numbers, letters, special symbols, and control codes. EBCDIC is unique to IBM and IBM-compatible equipment, primarily minicomputers and mainframes.

ASCII and EBCDIC are used in computers to communicate with peripheral equipment such as teletypewriters, video terminals, printers, and other input/output devices.

Go to the next frame.

Binary Arithmetic

---

34   Occasionally you will find it handy to be able to do binary arithmetic. Binary addition and subtraction are the most common operations. So let's take a look at the simple procedures for adding and subtracting binary numbers. If you can add and subtract decimal numbers, you have got it made.

The rules for binary addition are the same for addition of decimal numbers.

0	0	1	1	one plus
+0	+1	+0	+1	one equals
0	1	1	10	two

Note the binary 1 that occurs in the addition of 1+1. This is called the *carry*. We say "1 plus 1 equals zero with a carry of 1."

Here is an example of binary addition:

```
    1    ←  carry
 0011        3
+1010      +10
 1101       13
```

Now you try it. The binary sum of the decimal numbers 6 and 11 is:

    a. 0110
    b. 1011
    c. 10001
    d. 0001 0111

---

35   (*c.* 10001)

```
carries        11
            0110        6
           +1011      +11
           10001       17
```

Binary subtraction is just as easy since it too is just like decimal subtraction. The rules are:

			borrow ↓	
0	1	1	10	two minus
−0	−0	−1	−1	one equals
0	1	0	1	one

Since you can't subtract 1 from 0 in the fourth, or righthand, case in the above examples, you have to "borrow" a 1 from the next most significant bit to the left.

Here is an example:

```
 1011        11
-0110        -6
 0101         5
```

(continued next page)

Here is another:

```
 100110      38
-011101     -29
 001001       9
```

Here is one for you. The binary difference between the decimal numbers 37 and 30 is:

    a. 7
    b. 000111
    c. 001000
    d. 010001

---

36   (*b.* 000111)

```
 100101      37
-011110     -30
 000111       7
```

Sometimes it is necessary to represent negative quantities in binary form. In the decimal number system we put a minus sign in front of a number to tell us it is negative. The same is true in the binary number system:

$$-25 = -11001$$

However, we need a way to tell the computer that the number is negative. This is usually done by designating the MSB in a binary word as a sign bit. If this bit is a 0, the number represented by the remaining bits is positive. For example, +25 in the 8-bit byte below is:

```
MSB = Sign bit
 0   0   0   1   1   0   0   1  = +25
MSB                        LSB
```

If the sign bit is a binary 1, the number is negative. However, the remainder of the word is also changed to represent a negative number. This is done by creating the *complement* of the word. The complement of a binary number is formed by changing all 0s to 1s, and all 1s to 0s. For example, the complement of 1001 is 0110.

The complement of 101101 is:

    a. 010010
    b. 010011
    c. 110010
    d. 100101

---

37   (*a.* 010010)   Using this technique we express
the number −25 as:

        11100110

where the MSB is a 1 indicating a negative value and the remainder of the word is the complement of +25 or 0011001. An easy way to create a negative number is to simply write the positive value including the sign bit, then complement the result. Thus, a –37 is:

+37 = 00100101 (note the use of leading zeros)
–37 = 11011010 (complement)

The binary representation of the quantity –52 is:

    a. 11101000
    b. 11010011
    c. 11001011
    d. 11001100

---

38   (c. 11001011)

+52 = 00110100
–52 = 11001011

This representation of negative numbers is called the *ones (1s) complement.* It was once widely used in digital computers, but today most microcomputers use what is called *twos (2s) complement* representation of negative numbers. The 2s complement is formed by adding 1 to the LSB of the 1s complement. For example, the 2s complement of –52 is:

  11001011   1s complement
+00000001   add one to LSB
  11001100   2s complement

The 2s complement representation of –37 is:

    a. 00100110
    b. 10110101
    c. 00100100
    d. 11011011

---

39   (d. 11011011)   To convert a negative decimal number to 2s complement form, follow these steps:

    a. Convert the decimal number to binary.
    b. Write the binary number in its positive form showing the sign bit. Since all computers use a fixed word length, fill in all bits of the word. Use leading zeros where needed.
    c. Complement the number: change all 1s to 0s, and 0s to 1s, including the sign bit.
    d. Add 1 to the LSB.

(continued next page)

Express –75 in 2s complement form.

    a. 10101101
    b. 10110101
    c. 11001100
    d. 00011101

---

40   (*b.* 10110101)   To convert a 2s complement negative number to decimal, follow these steps:

    a. Subtract 1 from the LSB.
    b. Complement all bits including the sign bit.
    c. Convert to decimal.
    d. Add a minus sign.

The 2s complement number 10101011 in decimal is:

    a. –36
    b. –85
    c. –92
    d. –104

---

41   (*b.* –85)   It is important to note that using one bit to represent the sign of a number in a fixed-word-length computer reduces the maximum value a fixed-length word can represent. In an 8-bit word, the maximum value with no sign bit is $2^N – 1 = 2^8 – 1 = 256 – 1 = 255$. With a sign bit, only 7 bits remain to represent the value. Therefore, the maximum value positive number is 01111111 = 127. The maximum negative number is 10000000 or –128. Therefore, with an 8-bit word, the number representation range is –128 to +127.

What is the maximum positive and negative values you can represent with a 16-bit number?

    a. +4095, –4096
    b. +16,383, –16,384
    c. +32,767, –32,768
    d. +65,535, –65,536

---

42   (*c.* +32,767, –32,768)   With a full 16-bits, you can represent numbers as high as $2^{16} – 1 = 65,536 – 1 = 65,535$. If one bit is designated solely for the sign, that leaves 15 to represent the magnitude or $2^{15} – 1 = 32,768 – 1 = 32,767$. That is the maximum positive value. The maximum negative value is one more or –32,768 (1000 0000 0000 0000).

To confirm the above, what decimal value do you get if you assume 1000 0000 0000 0000 is a 2s complement number?

    a. –16,383
    b. –16,384

c. −32,767
d. −32,768

---

43   (d. −32,768)   All of the numbers we have discussed so far are whole numbers or integers. In the computer, we assume that the binary point is to the right of the LSB. But there are fractional numbers as well, such as 9.52 or .00136. How are such numbers represented in a computer? The answer is that the programmer takes care of it in the software. For example, if we multiply the two numbers above, assuming no decimal points, we would get 952 × 136 or 129,472. The programmer would write code that counts the number of decimal point positions to get the true answer .0129472.

We can also represent fractional numbers in a computer by assuming the binary point is to the left of the MSB. The weights of the positions are still some power of 2 but in this case a negative power of 2 to create fractional weights.

$$2^{-1} = 1/2 = .5$$
$$2^{-2} = 1/4 = .25$$
$$2^{-3} = 1/8 = .125$$
$$2^{-4} = 1/16 = .0625$$
$$2^{-5} = 1/32 = .03125$$
$$2^{-6} = 1/64 = .015625$$
$$2^{-7} = 1/128 = .0078125$$
$$2^{-8} = 1/256 = .00390625$$

Format of an 8-bit fractional number:

Binary point
↓

.	2^{-1}	2^{-2}	2^{-3}	2^{-4}	2^{-5}	2^{-6}	2^{-7}	2^{-8}
	MSB							LSB

The process of evaluating a fractional binary number is the same as with integers. For instance, the number 0.10011 is converted to decimal using the same process you used earlier. Identify those bit positions containing a binary and add their weights.

$$2^{-1} = .5$$
$$2^{-4} = .0625$$
$$+2^{-5} = .03125$$
$$\overline{\phantom{+2^{-5} = }.59375}$$

What is the decimal value of 0.010001?

   a. .25
   b. .26625
   c. .15275
   d. .06375

44   (*b*. 0.26625)   Go to the next frame.

Floating-Point Numbers

45   The numbers described in the previous frames are referred to as *fixed-point numbers* because the decimal points are fixed in place. It is also possible to represent numbers in binary using what is called *floating-point notation*. This permits very large and very small numbers to be represented.

Floating-point notation is based upon what we call *scientific notation*. This is the process of representing quantities by a number between 1 and 10 times some power of 10. For example, the number 10,000 is 1×10^4 or a 1 with 4 zeros appended to it. The value 5,273 is 5.273×10^3. Or, .000156 is 1.56×10^{-4}.

To change a number into its scientific notation format, move the decimal point to create a number between 1 and 10. Count the number of digit positions you moved the decimal. Multiply the number by a power of 10. Make the exponent + if you move the decimal point to the left and − if you move the decimal point to the right.

Convert .0000893 to scientific notation.

     a. 8.93×10^{-5}
     b. 8.93×10^{5}
     c. 8.93×10^{-4}
     d. $.893 \times 10^{6}$

46   (*a.* 8.93×10^{-5})   To create a floating-point number in binary, we do something similar. First, we convert numbers to some value between .1 and 1 multiplied by some power of 2. The exponent on the base 2 is the number of positions we move the binary point—plus to the left, minus to the right. The process is called *normalization*.

Assume the binary number 1100.11. To convert it to floating point, move the decimal point four places to the left, then add a power of two multiplier.

$$1100.11 = .110011 \times 2^4$$

The binary point is placed to the left of the MSB. Since we moved it four places to the left, we created a smaller number, so we must multiply by some positive power of 2 equal to 4. The resulting number is said to be *normalized*.

Another example is the value .001101. To normalize this value, you move the binary point to the left of the first binary 1 you encounter. To normalize a number you must always move the binary point so that the MSB is always a binary 1. Then you multiply the result by a power of 2 whose exponent is the number of positions moved. Since you moved the binary point to the right, you are creating a larger number, so you must multiply by a smaller factor. This requires a negative exponent. The result is the value $.1101 \times 2^{-2}$.

Determine the correct normalized values for 100001.001 and .000101.

    a. 1.00001001×2^5, 1.01×2^{-4}
    b. 1.00001001×2^6, 1.01×2^{-5}
    c. $.100001001 \times 2^5$, $.101 \times 2^3$
    d. $.100001001 \times 2^6$, $.101 \times 2^{-3}$

---

47   (*d.* $.100001001 \times 2^6$, $.101 \times 2^{-3}$)   To represent a floating-point number in the computer, you need two binary words, one for the value of the number and another for the value of the exponent. Both must have sign bits. A common arrangement is to use the format that follows.

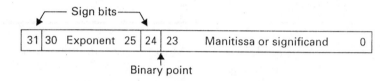

A 32-bit word, with the bits numbered from 0 to 31, is divided into two sections representing two 2s complement numbers. Bits 0 through 24 represent the value of the number in fractional form. Bit 24 is the sign bit and the binary point is assumed to be between bits 23 and 24. We call this the *mantissa* or *significand*.

The exponent on the power of 2 is represented by bits 25 through 31. Bit 31 is the sign bit while the value is represented by bits 25 through 30 or 6-bits. The exponent is assumed to be an integer. In this case, some whole number from 0 to $(2^6 - 1) = 64 - 1 = 63$.

Given this arrangement we can represent any value between about 2^{-64} to 2^{+63} with 24-bit precision. Now you can see the value of floating point.

The numerical value of a floating-point number without the binary point is determined by which part of the floating-point number?

    a. exponent
    b. mantissa
    c. mantissa+exponent
    d. sign bits

---

48   (*b.* mantissa)   Now you will learn two different ways to convert a binary floating-point number into its equivalent decimal number. Assume that the mantissa is +.10001000000000000000000 and the exponent is

(continued next page)

+000011. The mantissa portion is obtained by adding up the weights of those positions containing binary 1s, or in this case, .5 + .03125 or .53125. The exponent is $2^3 = 8$. The decimal equivalent is .53125 × 8 = 4.25.

Another way to make the conversion is to move the binary point the number of places designated by the exponent—in this case three places to the right—and then convert to decimal. The original mantissa is +.10001. The following zeros do not change the value. Moving the binary point three places to the right gives 100.01, which also converts to 4.25.

Assume the exponent is a negative value with a sign bit of 1 and a value of 111111. Treating it as a 2's complement number, how many places and in what direction must the binary point of the mantissa be moved?

    a. one place left
    b. one place right
    c. 63 places right
    d. 63 places left

49

(*a.* one place left)   Most microcomputers work well with fixed-point numbers and do not have the facilities to deal with floating-point numbers. Floating point is most useful for scientific, engineering, and higher math applications where serious "number crunching" is needed. It is not required in most ordinary business calculations. A typical microprocessor cannot handle floating point, but many of them are designed to work with an accessory floating-point integrated circuit (IC) called a math coprocessor. Many of the newer larger microprocessors such as Intel's Pentium and Motorola's 603 and 604 Power PC incorporate an on-chip floating-point arithmetic unit.

What is the name of the auxiliary chip that adds floating-point math to a microprocessor?

    a. math chip
    b. floating-point IC
    c. math ALU
    d. coprocessor

50

(*d.* coprocessor)   Now answer the Self-Test Review Questions on the diskette before going on to the next unit.

Computer Memories

Introduction to Memory

1   The computer memory is that part of the digital computer where data and the instructions making up a program are stored. There are two basic types of memory used in computers, RAM and ROM. *RAM* means *random access memory,* but it also refers to what we call *read/write memory.* This means that you can store data in it or retrieve data from it under program control. RAM is reusable and reconfigurable. Any programs or data in RAM can be written over if desired. RAM is also that part of memory where the program you are currently running is stored.

(continued next page)

The other type of memory is *ROM* or *read-only memory.*
You can only retrieve data from ROM under program control. You cannot write data to memory or store new information in it. Programs and/or data are initially stored in ROM when it is made, or later with the aid of a special piece of equipment called a programmer. The contents typically cannot be changed. The main virtue of ROM is that when the power to the computer is turned off, ROM retains its contents, whereas RAM loses all its data and programs.

Read/write memory is known as:

    a. ROM
    b. disk drive
    c. RAM
    d. RWM

---

2   (*c.* RAM)   All computers have both RAM and ROM. The amount or size of RAM can often be changed. For example, it is usually possible to add RAM to your personal computer if you need to work with larger programs or process greater volumes of data. Computer memory uses semiconductor circuits to store the data and instructions of a program in binary form. Semiconductor memory is made up of many individual storage cells, each capable of remembering one bit of information. Each storage cell can assume either of two states so that either a binary 0 or binary 1 can be stored. Different types of storage cells are used for RAM and ROM. We will examine RAM types in the next section and ROM later.

Figure 3-1 shows the various types of RAM and ROM used in computers. Virtually all computer memory, regardless of type, is packaged in integrated circuit form. In this unit you will learn about those memory ICs described in Figure 3-1.

Go to the next frame.

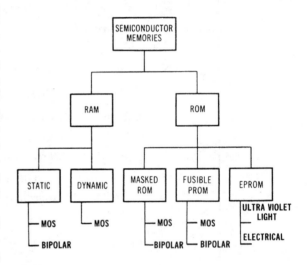

Fig. 3-1. Hierarchy of semiconductor memories.

RAM

---

3   There are two basic types of semiconductor RAM storage elements: *static* and *dynamic.* In both types of memories, the main component is the metal-oxide silicon field-effect transistor (MOSFET). *Static memory* uses a flip-flop as the storage element. A flip-flop is a digital circuit that can assume either of two states, set or reset. When the flip-flop is set, it is said to be storing a binary 1. When it is reset, it is storing a binary 0.

A typical static memory element is shown in Figure 3-2. The flip-flop circuit is made up of four MOSFETs. The states

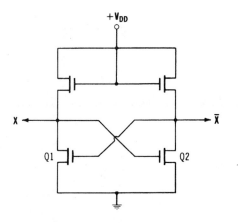

Fig. 3-2. MOSFET flip-flop static storage element.

of transistors Q1 and Q2 determine whether a 1 or a 0 is stored. In one state, Q1 conducts and Q2 is cut off. In the other state, Q2 conducts and Q1 is cut off. The other two MOSFETs are connected to work as load or drain resistors. The state of the flip-flop can be changed by pulling either the X or X NOT lines low with external circuits. The state of the flip-flop is determined by monitoring the X and X NOT lines. The X and X NOT outputs are always complementary; that is, when X = 1, X NOT = 0 and vice versa.

If the static flip-flop is set, the bit stored is:

    a. binary 0
    b. binary 1

4   (*a.* binary 1)   Static memory is used primarily in applications requiring small size and very high speed. A common application is cache memory. A *cache memory* is a small high-speed memory used between the CPU and the main computer RAM. Its purpose is to reduce the time it takes to access data and instructions from main memory. The result is a considerable increase in the speed of execution of programs and the access of data.

Another type of semiconductor storage element is called a *dynamic memory cell.* The dynamic memory cell is an integrated circuit capacitor. A *capacitor* is an electronic component that stores electrical energy in the form of an electric force field. Applying a voltage to a capacitor charges it. The capacitor will retain the charge until it is used, at which time the capacitor is discharged.

Energy stored in a capacitor is called a:

    a. MOSFET
    b. magnetic field
    c. voltage
    d. charge

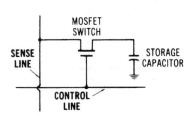

Fig. 3-3. Dynamic memory storage cell.

5   (*d.* charge)   A typical dynamic memory storage cell is shown in Figure 3-3. The basic storage cell is a capacitor. This is an integrated capacitor connected in series with a control transistor (MOSFET) that acts as an on-off switch. To store data in or read data from the memory cell, the MOSFET is turned on by applying a positive pulse to the control line. When the MOSFET is turned off, the capacitor is isolated from the rest of the circuitry. The state of the storage element is retained in the charge on the capacitor. When the capacitor is charged, a binary 1 is stored. When the capacitor is discharged, a binary 0 is stored. To read data from the cell, the MOSFET is turned on connecting the capacitor to the sense line. External circuits examine the state of the capacitor charge and send it to the CPU.

In Figure 3-3, the MOSFET is a:

(continued next page)

a. switch
b. storage cell
c. capacitor
d. chip

---

6 (*a.* switch)   Because the capacitance in a dynamic cell is very small, any charge on it will quickly leak off. Of course, this is undesirable since the state of the cell will change and data stored there will be lost. To overcome this problem, the charge on the cell must be "refreshed" periodically. Special "refresh" support circuitry is used to continuously restore the charge on dynamic memory cells. This refresh operation occurs approximately every two to four milliseconds in modern dynamic memories.

The process of renewing the charge on a capacitive cell to prevent loss of data is called:

a. restore
b. refresh
c. recharge
d. rejuvenate

---

7 (*b.* refresh)   Dynamic memory cells require more circuitry and are more complex in their operation than static cells because of the need for refresh. But the dynamic cells are smaller and many more of them can be contained on a given size silicon chip than static cells. As a result, dynamic memories cost less per bit than static memories. In addition, dynamic cells usually operate slower than static cells but consume less power. Dynamic RAM chips are used as the main memory in most personal computers because of their low cost given their massive amount of storage capacity.

Which statement is *not* true about dynamic RAM cells?

a. smaller than static RAM
b. less power than static RAM
c. faster than static RAM
d. less expensive than static RAM

---

8 (*c.* faster than static RAM)   On the other hand, static cells are simpler to use and require less external support circuitry. They are also faster at reading and writing. Bipolar (TTL, ECL) transistors rather than MOS are also used to make static memories that are considerably faster than memories made with MOSFETs.

An important point to remember is that any stored data is lost if power is removed from a memory cell. Power must be continuously applied if the cell is to retain the data. This applies to both static and dynamic cells. Memory cells with this characteristic are said to be *volatile.* A *nonvolatile* memory cell retains data even if power is removed; an example of a nonvolatile cell is the storage cell used in a ROM.

Both static and dynamic memory cells will lose data if the power is removed.

    a. True
    b. False

---

9   (*a.* True)   Go to the next frame.

Memory Organization

---

10   Many individual storage cells are made on a single silicon chip to form a memory integrated circuit (IC). You will hear these referred to as *RAM chips.* Dynamic RAM is called *DRAM,* while static RAM is called *SRAM.* Memory ICs are available with storage capacities of a few thousand bits to many millions of bits. A billion-bit chip is already designed, and commercial ICs should be available by the year 2000. As you can see, a tremendous amount of data can be stored in a single memory IC. A number of these ICs are then interconnected to form a complete memory for a microcomputer.

Memory IC chip capacities are typically some power of 2. Some common sizes are given in the table below:

Table 3-1. The Most Commonly Used Memory Sizes

Number of address bits	Number of memory cells
N*	2^{N}**
10	1,024
11	2,048
12	4,096
13	8,192
14	16,384
15	32,768
16	65,536
17	131,072
18	262,144
19	524,288
20	1,048,576
21	2,097,152
22	4,194,304
23	16,777,216
24	33,554,432
25	67,108,864
30	1,073,741,824
32	4,294,967,296

*N = Some power of 2. The number of bits in an address word that individually identifies each storage location in the memory.

**2^{N} = number of bit storage cells

(continued next page)

Dynamic RAM sizes go all the way to over 64 million bits. Static RAM sizes go up to about 16 million bits.

Each of the storage locations for a bit in a memory is identified by a binary word called the *address*. To access a bit or store a bit, the address is supplied to the memory that activates the desired location. The number of bits in the address word sets the maximum size of the memory. Table 3-1 shows the memory capacity for different address bit lengths (N).

If the address word has 8-bits, the RAM size is:

  a. 128 bits
  b. 256 bits
  c. 572 bits
  d. 768 bits

---

11 (*b.* 256 bits)   Because of the large numbers involved in denoting memory, shorthand has been developed to represent and discuss its capacity. For example, the letter K is used as a multiplier of 1,024. (In electronics the letter K usually means 1,000, but not in computers.) The letter K is attached as a suffix to a number to tell the memory size. The designation 64K represents memory with a capacity of 64 × 1,024 = 65,536 bits.

For larger memory, the suffix M is used to represent a multiplier of 1,048,576. (In electronics, M usually means 1,000,000. When discussing memory sizes, M means 1,048,576.) Therefore, memory of 32M means 32 × 1,048,576 = 33,554,432 bits.

When the designation is 32M *bits,* the abbreviation 32Mb is used. A lower case "b" represents bits. You will also see the capital letter B used to represent bytes. The designation 32MB means 32 megabytes, or 32 × 1,048,576 × 8 = 268,435,456 bits. (Remember that a byte contains 8-bits.) This could also be stated as 256Mb.

How many bits are contained in a 4MB RAM?

  a. 4,194,304
  b. 16,777,216
  c. 33,554,432
  d. 67,108,864

---

12 (*c.* 33,554,432)   Because dynamic cells are simpler, they are smaller and, as a result, more of them can be made on a given size silicon chip. The higher density makes dynamic memory more economical than static memory. Today, a 16Mb chip (16,777,216 bits) is the largest dynamic RAM chip currently used. Commercial 64Mb (67,108,864-bit) DRAM chips are already available, and larger, 256Mb and 1 Gb, chips are being developed (Mb = megabit, Gb = gigabit).

Regardless of the type of memory, the cells are arranged in various configurations. For example, a 1,048,576-bit memory IC can be organized as 1,048,576 1-bit words (1,048,576 × 1), as 262,144 4-bit words (262,144 × 4), as 131,072 8-bit words, or as 65,536 16-bit words. Note that all have the same number of bits (1,048,576), but each is organized differently.

How many bytes can a 4,194,304-bit RAM chip store?

> *a.* 65,536
> *b.* 131,072
> *c.* 262,144
> *d.* 524,288

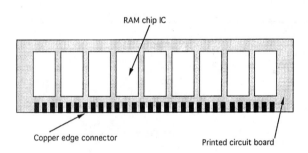

RAM chip IC

Copper edge connector

Printed circuit board

Fig. 3-4. A single in-line memory module (SIMM) may have 30 pins, as shown here, or 72 pins.

13

(*d.* 524,288)   Memory chips are usually mounted on a printed-circuit board to form larger memories. They are connected in arrays to form many storage locations for fixed-length, multi-bit words. For example, to store a byte (code or number), 8 storage cells are grouped together and treated as a single memory location. Common word sizes are 8, 9, 16, 32, and 64 bits.

DRAM ICs are frequently packaged on a thin PC board about the size of a stick of gum to form a fixed memory size. These are referred to as *single in-line memory modules* (*SIMMs*). See Figure 3-4. Each SIMM has a copper-edge connector that plugs into a slot on a larger socket capable of holding four, eight, or more SIMMs. These SIMM sockets are in turn mounted on a larger printed circuit board, usually the motherboard of a personal computer. SIMMs come in various sizes, such as 4M × 9 or 16M × 8.

A small PC board that holds multiple RAM ICs is called a:

> *a.* DRAM
> *b.* chip
> *c.* SIMM
> *d.* motherboard

14

(*c.* SIMM)   Each memory location can be loaded with data from an outside source (CPU, I/O device, etc.). We say that data can be stored in or "written" into the memory location.

Data can also be retrieved from memory to use elsewhere (CPU, I/O, etc.). We say that data is "read" from memory. Reading data from memory does not destroy it.

In order to identify the memory location used in a read or write operation, we must be able to distinguish it from all other locations. This is done by assigning each storage

(continued next page)

location a unique number called an address, as mentioned earlier.

To prepare a memory location for a read or write operation, it is first addressed. We do this by applying the binary equivalent of the address to the memory. This binary address enables or activates the desired memory location. This is called accessing the memory location.

The address is a fixed-length binary word. For example, an 8-bit word may be used to identify a number of unique storage locations for binary words. With 8 bits, a total of 2^8 = 256 states or numbers can be represented. Therefore, an 8-bit word can address or identify 256 storage locations. A 16-bit address can access 2^{16} = 65,536 words. A 24-bit address can address 2^{24} = 16,777,216 words.

The size of the address word determines the maximum memory size. The number N or the power of 2 in Table 3-1 is the number of address bits.

With 20 bits, how many memory locations can be addressed?

  a. 65,536
  b. 262,144
  c. 1,048,576
  d. 4,194,304

---

15 (c. 1,048,576 or 1M)   Since the address is given in binary, the memory size is always some power of 2. Table 3-1 shows the most commonly used memory sizes.

As memories get larger, larger designators are needed. Already billion bit and billion byte memories are available. The designator G for "giga" for billion is widely used. One giga (usually pronounced "one gig") means 1,073,741,824 bits, bytes, or other size words.

A 1GB RAM stores how many total bits?

  a. 1,073,741,824
  b. 2,147,483,648
  c. 4,294,967,296
  d. 8,589,934,592

---

16 (d. 8,589,934,592)   A block diagram of a typical microcomputer memory is shown in Figure 3-5. Note that with a 10-bit address word, a total of 2^{10} = 1,024 words can be addressed. The memory locations are numbered 0 through 1,023. An 8-bit word can be stored in each location. Eight data lines are used to store or retrieve a word from memory. Note that the address is applied to an address decoder. This is a logic circuit that looks at the 10-bit binary address, then identifies and enables the single location corresponding to that address.

Refer to the memory block diagram in Figure 3-5. Assume that the binary number 0000001001 is applied to the

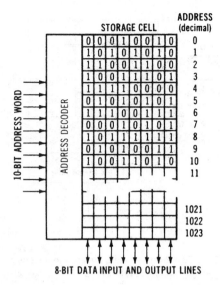

ADDRESS (decimal)
STORAGE CELL

0 0 0 1 0 0 1 0	0
1 0 1 0 1 0 1 0	1
1 1 0 0 1 1 0 0	2
1 0 0 1 1 1 0 1	3
1 1 1 1 0 0 0 0	4
0 1 0 1 0 1 0 1	5
1 1 1 0 0 1 1 1	6
0 0 0 1 0 1 0 1	7
1 0 1 1 1 1 1 1	8
0 1 0 1 0 0 1 1	9
1 0 0 1 1 0 1 0	10
	11
	1021
	1022
	1023

10-BIT ADDRESS WORD

ADDRESS DECODER

8-BIT DATA INPUT AND OUTPUT LINES

Fig. 3-5. Block diagram of a random-access memory.

address input and a read operation is specified. The binary output of the memory will be:

a. 00001001
b. 01010011
c. 10010001
d. 11001010

17 (*b.* 01010011) The address 0000001001 (decimal 9) enables location 9 and the word stored there, 01010011, is placed on the output lines.

To store the word 11101101_2 in location 11_{10}, the address 0000001011 is applied to the memory. The address 0000001011 is the binary equivalent of the decimal location 11.

Referring to the memory in Figure 3-5, you can determine its total word storage capacity by noting the number of memory locations and the number of bits stored in each location (address). There are 1,024 locations and eight bits per location. Using the shorthand designation explained earlier, this memory can store 1KB or 1K × 8 words. The total number of bits stored is 1,024 × 8 = 8,192.

Remember that the number of bits in the address word determines the maximum number of locations that the computer can address. Also, the number of address bits is the number N used on the power of 2 to determine memory size.

How many address bits must be used to address 4MB? (Use Table 3-1).

a. 18
b. 20
c. 22
d. 24

18 (*c.* 22) The memory organization diagram in Figure 3-5 shows how memory looks to the user or programmer; that is, there are sequentially numbered locations for fixed-length words. In reality, the hardware of the

(continued next page)

RAM is typically much different. For example, most RAM ICs are internally organized as a matrix. Figure 3-6 illustrates this concept with a small number of storage cells. The storage cells are arranged in rows and columns. Here a 16-bit memory is arranged as four rows and four columns. A 4-bit address is needed to address all sixteen locations. The addresses are 0000 through 1111. This is divided into two 2-bit segments. One segment feeds a 2-bit to 4 line decoder that selects one of the four rows. The second 2-bit segment of the address drives another 2-bit to 4 line decoder that selects one of the four columns. At the intersection of the selected row and column is a single memory cell that is enabled to perform a read or write operation. Figure 3-6 shows how the location 0110 is enabled.

A 1 Kb RAM chip is usually organized as a 32 × 32 matrix for a total of 32 × 32 = 1,024 locations. A 10-bit address is needed to address all locations. This is divided into two 5-bit segments. One segment feeds a 5-bit to 32 line decoder that selects one of the 32 rows. The second 5-bit segment of the address drives a 5-bit to 32 line decoder that selects one of the 32 columns. At the intersection of the selected row and column is a single memory cell that is enabled to perform a read or write operation.

The organization of larger RAM chips is the same. A 1 Mb RAM chip is arranged as 1,024 horizontal rows and 1,024 vertical columns, giving 1,024 × 1,024 = 1,048,576 total bit locations. To address 1 Mb requires a 20-bit address word. This word is divided into two 10-bit words. For a given 20-bit address input, one 10-bit to 1,024 decoder enables one of the 1,024 rows and another 10-bit to 1,024 decoder enables one of the 1,024 columns. At the intersection of the enabled rows and columns is a one-bit cell that is selected for a read or write operation.

Most RAM IC chips have a limited number of connection pins (usually 16 or 20). There is usually an insufficient number of pins to accommodate the entire address word. To overcome this problem, the address word is usually multiplexed in two parts onto the address pins. A 4 Mb RAM chip needs a 22-bit address word but only has 11 address pins. This address is applied in two steps. One 11-bit half is applied to the address pins and it is stored on-chip in an address register. The second half of the address is then applied to the same 11 pins and it too is stored in the register. Upon receiving the second part of the address, the enabled bit is then ready to read or write.

In a 1 Mb RAM chip organized as 1,048,576 × 1, only one of the cells is enabled. To store 1 MB, or one megabyte, eight of the 1 Mb RAM chips are needed. The same address is applied simultaneously to all chips. Therefore, one bit in each chip is selected for each address. In other words, each chip stores one bit of the 8-bit word.

Figure 3-7 illustrates the idea with smaller memories. The 16 × 4 memory can store sixteen 4-bit words. Note that four 16-bit RAMs are used and each stores one bit of each 4-bit word.

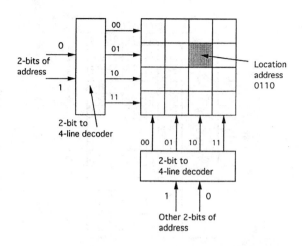

Fig. 3-6. Two 2-bit decoders are used to decode a 4-bit address capable of addressing 16 memory cells.

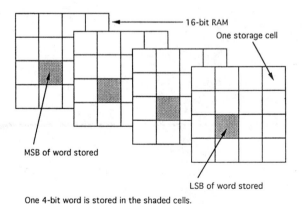

One 4-bit word is stored in the shaded cells.

Fig. 3-7. One 4-bit word is stored in the shaded cells.

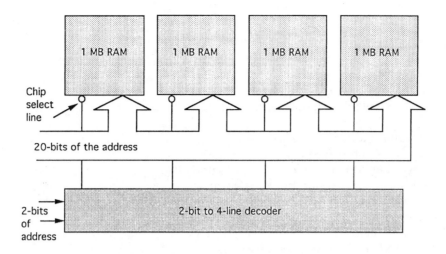

Fig. 3-8. A 4 Mb RAM uses four 1 Mb chips. All chips receive the 20 least significant bits. The two most significant bits are separately decoded to enable one of the four 1 Mb chips via the chip select lines.

To create larger memories, multiple segments are grouped together. For instance, a 4 MB RAM can be created by combining four of the 1 MB memories just described. See Figure 3-8. A 22-bit address is needed to address all four megabytes. The 20 least significant bits (LSBs) of the address are applied simultaneously to all four $1M \times 8$ groups. The remaining two most significant bits (MSBs) are applied to a 2-to-4 line decoder. The four outputs go to enable one of the four 1 MB groups. The two MSBs select one of the four groups; the other three are disabled. All RAM chips have a select or enable line that turns on the chip. If the select line goes low, the chip is selected for a read or write operation. If the select line is high, the chip is disabled. In each 1 MB memory group, all of the IC chip select lines are tied together and to one output of the 2-to-4 line decoder.

A RAM chip is organized as a 512×512 matrix. How many of these RAM chips would be needed to create a 2 Mb × 16 RAM?

    a. 16
    b. 32
    c. 64
    d. 128

19   (*d.* 128)   A 512×512 matrix produces a total of 262,144 storage locations or a 256K × 1 organization. It takes four of these chips to store 1Mb or eight chips to store 2 Mb. With these eight chips you can store 2M or 2,097,152 one-bit words. To store that many 16-bit words requires a total of $8 \times 16 = 128$ chips.

(continued next page)

Once the memory location is enabled by applying the desired address, data may be written into or read from that location. The address is usually supplied by the CPU. Only the addressed memory location is enabled. The data stored in the enabled location usually, but not always, comes from or goes to the CPU. Remember, a microprocessor is a single-chip CPU.

An important specification for a RAM is its speed of operation or the *access time.* Access time is the time it takes for a word to be addressed and the RAM location prepared for a read or write operation. Remember that DRAMs must be repeatedly refreshed or the data is lost. During the refresh time a read or write operation cannot be performed. The refresh cycle interferes with the access to the DRAM, which produces delays and slows down overall computer operation.

Typical access times for dynamic RAMs are in the 50 to 250 nanosecond (nS) range. DRAMs of 60 or 70 nS are very common. The faster RAMs are preferred so they can keep up with the microprocessor, which is usually much faster.

Static RAMs are even faster. They have an access time in the 5 to 50 nS range. Most SRAM are CMOS with access times of 20 to 50 nS. The 5 to 20 nS access times are achieved with SRAMs made with ECL logic circuits.

Select the *most correct* answer.

    a. Capacitive storage elements are inferior to flip-flops.
    b. DRAMs are faster than SRAMs.
    c. DRAMs are most commonly used in the main RAM of a PC.
    d. The microprocessor determines the access time.

---

20   (*c.* DRAMs are most commonly used in the main RAM of a PC.)   Go to Frame 21.

Read-Only Memories

---

21   Another kind of memory widely used in microcomputers is the *read-only memory,* or *ROM.* Some types of ROMs have the data and instruction words permanently stored in them when they are manufactured. Such memories are usually called *masked ROMs.* In other types of ROMs, the data or instructions are semi-permanently stored in them after they are manufactured. ROMs are also random access memories in that any location may be addressed at random at any time, but the data may only be retrieved. Data cannot usually be stored or written into the memory under the control of the CPU.

Most microcomputers use a combination of RAM and ROM, but ROM represents only a small part of the total memory used. Most memory in personal computers, for instance, is DRAM. The desired program and data is loaded into DRAM from a disk drive each time power is applied. Personal computers also contain a small ROM that stores programs to start (boot) and test the computer when it is turned on. Other programs in ROM are those that operate the keyboard and video monitor.

ROMs make up a larger part of the total memory complement in embedded microcomputers dedicated to a fixed application. The desired program and data are permanently stored in the ROM. Therefore, the microcomputer always executes the same set of instructions and performs the same function.

The CPU cannot write data into a ROM.

> *a.* True
> *b.* False

---

22   (*a.* True)   The major advantage of a ROM over a RAM or read/write memory is that the ROM is nonvolatile. When power is removed from a ROM, its contents remain undisturbed. However, all programs and data in DRAM or SRAM are lost.

Another common application of ROMs is as mass storage media, such as a floppy disk. The ROM is installed in a removable or plugable program cartridge. It is mounted on a printed circuit board and housed in a low-cost plastic container. The result is a plug-in mass-storage module. This storage module is designed so that it can be plugged into the bus of a microcomputer system. This technique is widely used in programmable video games and to store type fonts in laser printers. With a system of plug-in memory modules, the function or application of the computer is quickly changed by simply plugging in a different module. Access to data sources is as convenient with ROM cartridges as it is with floppy disks.

Keep in mind that all ROMs are also random-access memories because any word can be accessed at random by applying the desired address. But ROMs are not called RAMs. The term RAM is only used to refer to read/write memories.

The term that best describes the main feature of a ROM is:

> *a.* random access
> *b.* temporary
> *c.* volatile
> *d.* nonvolatile

23   (d. nonvolatile)   While data is initially stored in a ROM during its manufacture, other types of ROMs permit the user to store the data in them. These ROMs are referred to as *programmable read-only memories,* or *PROMs.*

An instrument called a PROM programmer is used to load data and instruction words into the desired locations. The PROM programmer is usually connected to a personal computer. The desired data or program is usually created on or stored in the PC and then transferred to the programmer that puts the data into the PROM.

In one kind of PROM, tiny fuses are used in each storage cell to determine if a 1 or 0 is to be stored. The PROM programmer "blows" fuses where a binary 1 is wanted. For a

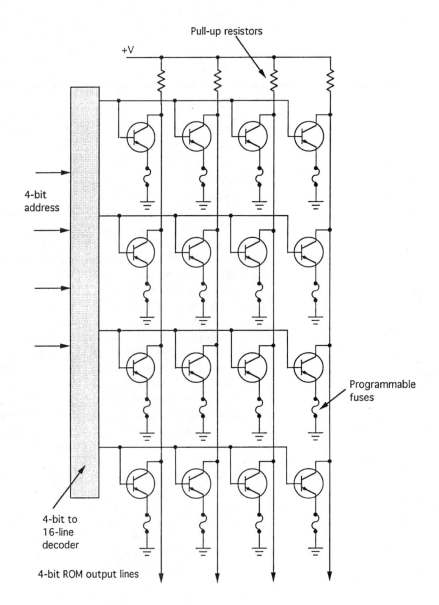

Fig. 3-9. A 16-bit ROM made with a matrix of bipolar transistors with fuses in the emitter leads.

Floating gate prevents conduction
if it is negatively charged.

Drain

Gate

MOSFET conducts from source to drain
if a positive voltage is applied to the gate.

Source

Schematic symbol of a floating gate MOSFET.

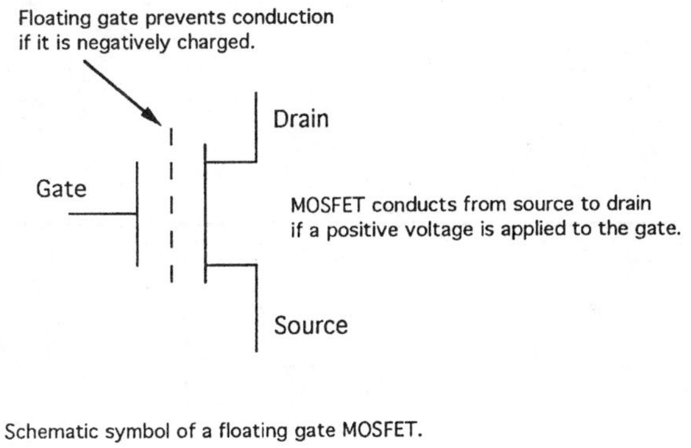

Fig. 3-10. Schematic symbol of a floating gate MOSFET.

binary 0, the fuses are left alone. Such PROMs are referred to as fusible PROMs. Once programmed or "burned," a PROM cannot be altered because the storage is permanent.

Figure 3-9 shows simple 4-bit four-word ROM. A 4-bit address word drives a one-of-four decoder that selects one of the four words. The selected decoder output line goes high and will turn on those transistors whose fuses in the emitter circuit are good. The transistors will conduct and bring the output lines low (binary 0). If the fuse is blown, the transistor will not conduct even if its base is high, since the emitter is open. Therefore, the pull-up resistor makes the output line high (binary 1). While bipolar transistors are shown in Figure 3-9, MOSFETs can also be used.

Another widely used PROM is the erasable PROM or EPROM. The storage cell in this type of PROM is a MOSFET with a floating gate that is positioned between the normal gate element and the other elements. The schematic symbol for a floating gate MOSFET is shown in Figure 3-10. The MOSFET acts as a switch that is either cut-off or conducting. The MOSFET conducts if you apply a positive voltage to the gate element. The floating gate can be programmed by storing a negative charge on it. This prevents the positive gate voltage from turning on the MOSFET. Once charged, the floating gate can retain the charge for years even when power is removed from the circuit. The PROM is set up so that a charged floating gate produces a binary 0 while an uncharged floating gate allows the MOSFET to turn on and produce a binary 1.

The contents of the EPROM are erased by exposing it to ultraviolet light. This light passes through a quartz window on the IC and exposes the silicon chip that contains the memory cells that are erased. The ultraviolet light discharges the floating gates in all MOSFETs.

PROMs and EPROMs are often said to be "field programmable" because they are programmed in the field by the customer or user rather than at the factory. A PROM programmer is used to write the new data into the ROM. But unlike fuse PROMs, the EPROM can be erased and reprogrammed.

A PROM programmer is used to:

a. read data from a PROM
b. write instructions and data into a PROM
c. erase a PROM
d. correct problems and errors in a PROM

24   (b. write instructions and data into a PROM)
Another form of EPROM is the EEPROM or electrically erasable PROM. The EEPROM features circuitry that will erase all of the memory content upon the application of an external voltage pulse. Then it can be reprogrammed. The basic storage element is the floating gate MOSFET.

(continued next page)

The erase process is relatively slow, even slower than the refresh cycle in a DRAM. Writing data into EEPROM is also relatively slow, much slower than the access time of a DRAM. Nevertheless, some of the newer EEPROMs can be written into by the CPU under program control. The read access time of an EPROM, however, is very fast, typically in the 10 to 50 nS range.

A new class of EEPROM is the flash PROM or flash memory. The term *flash* is used because the erase, write, and read access times are extremely fast, approaching the access times of a DRAM. And like all ROMs, they are non-volatile. Flash ROMs can be used as RAM but they are much more expensive than conventional DRAM. However, flash PROMs are showing up as mass storage devices that can replace disk drives in some applications like laptop/notebook computers. A flash PROM is:

 a. a fast EEPROM
 b. a type of SRAM
 c. a type of DRAM
 d. ultra violet erasable

25

(*a.* a fast EEPROM)   EPROMs are most widely used in embedded or dedicated micros. They are available as separate products, but also some single-chip microcomputers have a small built-in EPROM or EEPROM.

Programs to be stored in an EEPROM are first written on a personal computer and stored on a disk and in RAM. The PROM programmer is connected to the PC. The microcomputer containing the EEPROM is plugged into the programmer socket. It is then erased by applying an erase pulse. Then the program to be stored in the EEPROM is downloaded from the PC to the programmer that writes the program into the EEPROM circuits. The EEPROM or microcomputer containing the EEPROM is then installed in the final product.

The source of a program to be stored in an EPROM is a

 a. PROM programmer
 b. personal computer
 c. a PROM cartridge
 d. a CD-ROM

26

(*b.* personal computer)   Now answer the Self-Test Review Questions on the diskette before going on to the next unit.

How Microprocessors Work

LEARNING OBJECTIVES

When you complete this unit, you will be able to:

1. Define the terms *register, accumulator, general-purpose register, program counter, instruction register, memory address register, op code, data bus, parallel data transfer,* and *bidirectional data transfer.*

2. Name the major registers in the 8080/8085, Z-80, and 8051 microprocessors.

3. Name and explain the various instruction-word formats used in popular microprocessors.

4. Explain the operation of a CPU as it executes a program.

CPU Architecture and Operation

1   As you learned previously, a microprocessor is an integrated-circuit central processing unit (CPU). When a CPU is combined with memory and I/O circuits, a complete microcomputer is formed. In this unit, you will see how these work together.

The main circuit element in a CPU is called a *register.* A register is a digital circuit capable of storing one, fixed-length, binary word. It is very much like a single memory location. All registers use flip-flops as the storage cells, with each flip-flop storing one bit of data. Data can be written into or read from a register. Usually the data to be processed by the CPU is taken from memory and stored temporarily in a register. See Figure 4-1.

Like a memory location, a register usually has a fixed length such as 8, 16, or 32 bits. But unlike a memory location, registers are often used to manipulate data as well as

(continued next page)

store it. That is, the register can alter or process the data in some way.

A register that can store four bytes contains how many flip-flops?

    a. 8
    b. 16
    c. 24
    d. 32

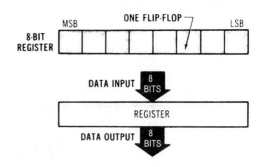

Fig. 4-1. Symbols used to represent a register.

---

2   (*d.* 32)   Here are some examples of how a register can process data. A binary word in a register can be shifted one or more bit positions to the right or left, or the register may be connected as an up/down counter so it can be incremented (add one to the content) or decremented (subtract one from the content). A register can also be reset or cleared, thereby erasing any data in it and leaving the content zero. Refer to Figure 4-2.

Data transfers and manipulations performed on a register are initiated by individual computer instructions. For example, one instruction may cause the register to be loaded from a memory location. Another instruction may cause the word in the register to be transferred to a memory location. See Figure 4-3. All data transfers are parallel, meaning that all bits are moved simultaneously from the source to the destination.

Register operations are initiated or specified by individual computer:

    a. software
    b. programs
    c. data words
    d. instructions

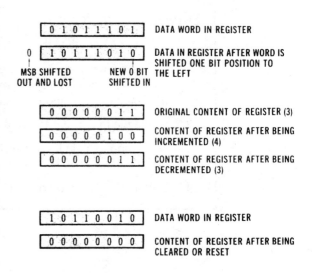

Fig. 4-2. Typical register operations.

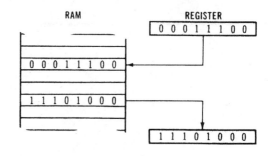

Fig. 4-3. RAM-to-register and register-to-RAM data transfers.

---

3   (*d.* instructions)   The instruction indicates the operation to be performed. This instruction is interpreted by the control section of the CPU that generates the timing signals that actually carry out the operation.

The main working register in most digital computers is called the *accumulator.* It may also be called the A register. The accumulator can hold one word whose bit length is equal to that of a memory location. An 8-bit microproces-

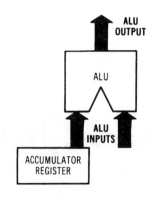

Fig. 4-4. Accumulator and ALU.

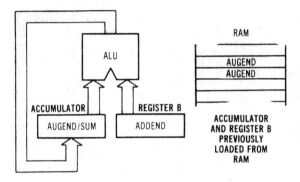

Fig. 4-5. Adding two numbers in the ALU.

sor has an 8-bit accumulator. The accumulator is part of the arithmetic/logic section.

The accumulator can be loaded from memory or the accumulator content can be stored in any memory location. Input/output (I/O) operations with peripheral units also usually take place through the accumulator.

Data to be processed by the arithmetic/logic section is usually held in the accumulator. This data is fed to the arithmetic/logic unit (ALU). The ALU is a digital logic circuit that can add, subtract, and perform a wide variety of logic operations. See Figure 4-4.

The ALU is capable of processing two inputs, one from the accumulator and one from the memory or another register. Referring to Figure 4-5, let's assume that we wish to add two binary numbers. To do this, we first load one of the numbers (the augend) into the accumulator. This is done with a load-accumulator instruction that takes the number from some memory location and puts it into the accumulator. Next, an "ADD" instruction is executed. This causes another word (the addend) to be taken from memory and put into register B. It is then added to the content of the accumulator. The sum is usually stored back in the accumulator, replacing the augend originally there.

The results of an arithmetic operation is usually stored in:

a. the B register
b. a memory location
c. the accumulator
d. ROM

4   (*c*. the accumulator)   Some microprocessors have two or more accumulator registers that share a single ALU. Two accumulators provide greater flexibility in the manipulation of data than a single accumulator. Such multiple registers simplify, speed up, and shorten a program to perform a given operation. Some sophisticated CPUs have four, eight, sixteen, or even more accumulators. Usually they are referred to as *general-purpose registers* (*GPRs*) or a *register file.* These registers can each use the ALU and act as temporary storage locations for data and the intermediate results of calculations. Instructions are provided to move data from one register to another.

In most microprocessors, the general-purpose registers share the single ALU. The ALU accepts only two input words and generates a single output word. The two words to be processed by the ALU usually come from any two GPRs or any GPR and a designated memory location. The destination of the ALU output can also be one of the GPRs in some CPUs.

Now let's look at a real microprocessor. Refer to Figure 4-6. This is a block diagram of the poplar 8080/8085A microprocessor.

(continued next page)

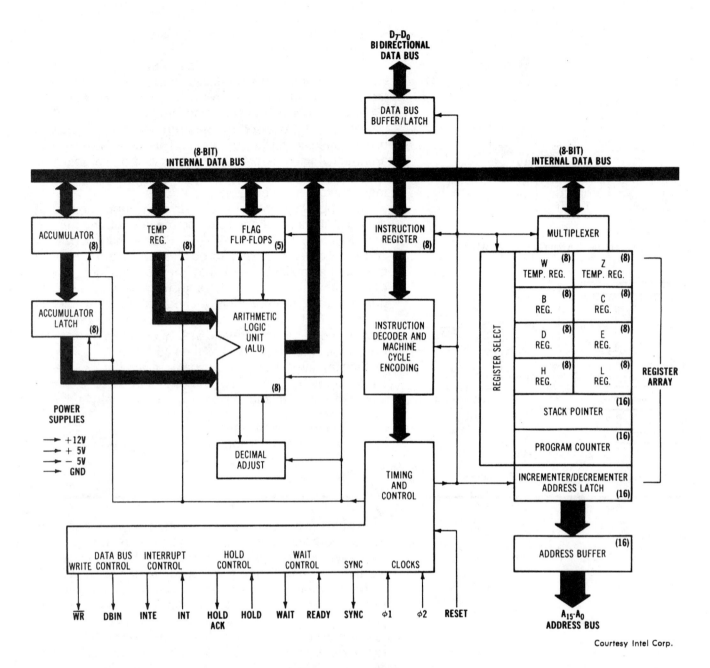

Fig. 4-6. Block diagram of the 8080/8085 CPU.

You may be wondering why we chose the older 8080/8085A microprocessor to use as an example. Well, first, although it was one of the first microprocessors developed, it was widely used in early personal computers and embedded controller applications. Literally hundreds of millions of these micros have been sold. Although it is no longer widely used for new designs, millions of products incorporating these devices are still in use.

In addition, several other micros have been spun off of this basic design. Some examples are the Z80 and its derivatives and the still popular 8051 series widely used for embedded applications. The design is simple, straightforward, or, as they say, classical. Above all, it is easy to learn. It doesn't matter what processor you learn initially, because all others are similar, as you will see in later chapters where other popular microprocessors are introduced.

Although there are some segments of Figure 4-6 that have not been explained yet, many will be clear to you. By the time you complete this unit and the next, you will fully understand not only the organization and operation of the 8080/8085 and Z80 but also all other microprocessors.

First, to get oriented, locate the accumulator and the ALU in Figure 4-6. The accumulator output feeds the accumulator latch. The latch is another 8-bit register that holds the content of the accumulator and feeds one input of the ALU. The other ALU input comes from a temporary (TEMP) register. The output of the ALU is fed back to the accumulator.

Refer to Figure 4-6. The main working register is the:

    a. accumulator
    b. latch
    c. temp register
    d. B register

---

5   (*a.* accumulator)   Now, referring again to Figure 4-6, note the register array shown on the right. Some of these registers are used for internal operations and cannot be used by the programmer. The W and Z temporary registers and the incrementer/decrementer address latch are examples. All of the other registers in this group are accessible by the programmer. Registers B, C, D, E, H, and L are each 8 bits in length and can be loaded, stored, incremented, decremented, or used in data transfers and arithmetic operations (i.e., add, subtract). Although not referred to as general-purpose registers, this is what they are.

You will note in Figure 4-6 that registers B, C, D, E, H, and L are grouped in pairs. Although they can each be used separately, they can also be used in pairs to form 16-bit registers. The 16-bit registers thus formed permit the storage and manipulation of 16-bit words. The three register pairs are labeled BC, DE, and HL.

CPUs also have several other registers besides the accumulator or GPRs. Some of these other registers are the *instruction register* (*IR*); the *program counter* (*PC*), also called the *instruction counter*; and the *memory address register* (*MAR*), also called the *address buffer*. Let's take a look at each of these registers.

First, locate and identify these registers in Figure 4-6. The *instruction register* is used to store the instruction word. When the CPU fetches an instruction from memory, it is temporarily stored in the IR. The *instruction* is a binary word or code that defines a specific operation to be per-

(continued next page)

formed. The instruction word is also called the *op code* or *operation code.* The CPU decodes the instruction, then executes it.

The size of the IR is:

    a. 4 bits
    b. 8 bits
    c. 16 bits
    d. 32 bits

---

6   (*b.* 8 bits)   The *program counter (PC)* is really a counter and a register. It stores a binary word that is used as the address for accessing the instructions in a program. If a program begins with an instruction stored in memory location 43, the PC is first loaded with the address 43. The address in the PC is applied to the memory, causing the instruction in location 43 to be fetched and executed. After the instruction is executed, the PC is incremented (add 1) to the next address in sequence, or 44. The instructions in a program are stored in sequential memory locations.

The *memory address register (MAR)* or *address buffer* also stores the address that references memory. This register directly drives the address bus and the memory address decoder in RAM or ROM. The MAR gets input from the PC when an instruction is to be accessed. See Figure 4-7. The MAR can also be loaded with an address that is used to access data words stored in memory. To retrieve a data word used in an arithmetic operation, the MAR is loaded with the binary word that points to the location of that word in RAM. This address is often a part of the instruction.

The MAR receives addresses from the:

    a. accumulator
    b. DE register
    c. program counter
    d. memory

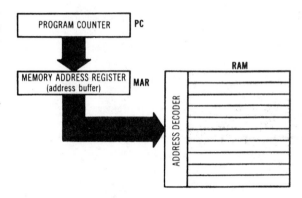

Fig. 4-7. Address flow for instructions in RAM.

---

7   (*c.* program counter)   You will see later how the instruction supplies an address to the MAR to locate desired data words in RAM.

It is important to note that the program counter and the MAR (address buffer) are 16-bit registers. This means that the address bus has 16-bits to address RAM and ROM. With 16 bits, a maximum of $2^{16} = 65,536$ words can be addressed.

To simplify the drawing of a microprocessor, usually, only those registers that can be referenced and used by the programmer are shown. An example is shown in Figure 4-8. Look it over and locate all those registers we have discussed. The MAR or address buffer is never shown.

In Figure 4-8, there are two registers that have not yet been discussed. These are the *flag* and *stack pointer registers.*

ACCUMULATOR (A)	FLAGS (F)
B	C
D	E
H	L
PROGRAM COUNTER (PC)	
STACK POINTER (SP)	

Fig. 4-8. Simplified register diagram of the 8080/8085 microprocessor.

The *flag* or *F register* is an 8-bit register whose individual flip-flops are set and reset by the ALU as the various arithmetic and logic operations are carried out. Each flip-flop is called a *flag*. As an example, there are zero (Z) and carry (C) flags. If the accumulator content is zero after an operation is performed, the Z flag is set indicating this condition. If an arithmetic operation (addition) results in a carry from the MSB of the accumulator, the C flag is set indicating this condition. These flags can be monitored or tested by the control circuitry to change the sequence of processing.

The *stack register* is a 16-bit register used to address a selected area of RAM known as the stack. This memory is used to store register contents and status information when subroutines and interrupts are used. You will learn more about the flags and the stack pointer in another unit.

If the accumulator is clear, the Z flag will be:

    a. binary 0
    b. binary 1
    c. either of the above
    d. can't tell

MAIN REGISTER SET		ALTERNATE REGISTER SET	
A	F	A′	F′
B	C	B′	C′
D	E	D′	E′
H	L	H′	L′

INDEX REGISTER X
INDEX REGISTER Y
PROGRAM COUNTER
STACK POINTER

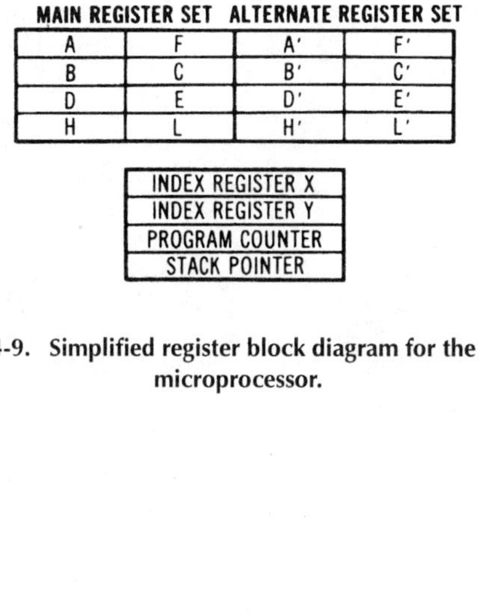

Fig. 4-9. Simplified register block diagram for the Z-80 microprocessor.

8 (*b.* binary 1) Another widely used microprocessor is the popular Z-80. It is basically an expanded, faster, and more powerful version of the 8080/8085. Its increased power and computing speed comes from the fact that it has more registers than the 8080/8085. Figure 4-9 shows the block diagram of the Z-80 register. Note that the main register set is exactly like that of the 8080/8085. However, the Z-80 has an alternate register set that is identical to the main register set. The whole group can be referred to as general-purpose registers. These extra registers speed up computer operations since fewer references to memory and fewer inter-register exchanges need to be made in any given computation. The Z-80 also has 16-bit program-counter and stack-pointer registers like the 8080/8085.

Refer to Figure 4-9. The Z-80 has two other 16-bit registers not contained in the 8080/8085. These are called *index registers*. Index registers are used in addressing data words. Their use will be explained in the next unit. The Z-80 is still used in some embedded applications. Numerous high speed dedicated microcontrollers are based upon the Z-80 architecture.

Which of the following registers are unique to the Z80 and not in the 8080/8085?
    a. accumulator
    b. index registers
    c. general-purpose registers
    d. program counter

9 (*b.* index registers) A microcontroller that has an architecture similar to the 8080/8085 is the highly popular 8051. The 8051 is a complete computer on a chip including an 8-bit CPU, RAM, EPROM, and I/O circuits. The CPU is a simplified version of the 8080/8085. It has only an 8-bit accumulator plus one other 8-bit register called the B register. It also contains various temporary registers, a status register, a stack pointer, and a program counter. It has two memory address registers that are used to separately address data and program memories. There are numerous versions of the 8051 with different memory and I/O configurations for different applications. We will discuss this chip in more detail later.

Instruction-Word Formats

10 There are three types of instruction formats used in typical 8-bit microprocessors such as the 8080/8085, Z-80, and 8051. These are illustrated in Figure 4-10. In the single-word format, the instruction is a single 8-bit word. This word is called the *op code*. The op code tells the ALU, the registers, and other elements of the system what to do. In this format, no address is used. The focus of the instruction is implied in the instruction. That is, the data to be processed is already in a location designated by the instruction. Usually the data is in a register. Typical instructions using this format are register-to-register transfer, shift data left (or right), or return.

The two-word instruction format in Figure 4-10 requires two 8-bit words to define the operation. These two words are stored in sequential memory locations. The first word is the op code. The second word is usually an address that specifies a memory location where the data word to be processed is stored. For example, if the op code calls for an "add" operation, the address word designates the location in RAM of the number to be added to the contents of the accumulator.

How many data words can be accessed with an 8-bit address word?

a. 8
b. 128
c. 256
d. 1,024

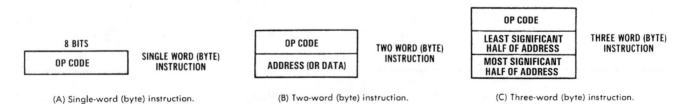

(A) Single-word (byte) instruction.   (B) Two-word (byte) instruction.   (C) Three-word (byte) instruction.

Fig. 4-10. Instruction-word formats.

11   (*c.* $2^8 = 256$)   The 256 bytes of RAM can be addressed with one byte of address.

In some two-byte instructions, the second byte is not the address. Instead, it is the data itself. This is called an *immediate instruction* since it is not necessary to address the data that is available immediately within the instruction itself.

The three-word instruction format in Figure 4-10 is made up of an 8-bit op code and two 8-bit address words stored in sequential memory locations. The second and third bytes together form a 16-bit address word that designates the location in RAM of the data to be processed. There are also three-byte immediate instructions where the second and third bytes are data words rather than an address.

In the three-byte instruction format, the first byte is the op code, the second byte is the least significant half of the address, while the third byte is the most significant half of the address. Several other formats will be discussed later when other microprocessors are described.

Suppose that 10111000 is stored in the second byte of an instruction and 11100101 is stored in the third byte. The hex address of the memory location where the data is stored is:

> *a.* 1DA7
> *b.* E5B8
> *c.* 7A1D
> *d.* 8B5E

12   (*b.* E5B8)   The total address, 1110010110111000, yields the hex equivalent E5B8 when divided into 4-bit groups.

To access a word in RAM, the instruction address word must be stored in the MAR. This happens during the instruction fetch operation. When an instruction is fetched from memory, the op code is stored in the instruction register while the address is stored in the MAR. The instruction is then executed.

The MAR usually gets its input from the program counter (PC). Once an instruction is fetched and executed, the PC is incremented.

The PC may be incremented once, twice, or three times depending on the length of the instruction just executed. If a two-byte instruction is executed, the PC is incremented twice so that the PC points to the address of the next instruction op code. See Figure 4-11. The program counter content is 10, so the instruction in location 10 will be fetched and executed. Because this is a two-byte instruction, the PC is incremented twice to 12 so that it points to the next op code.

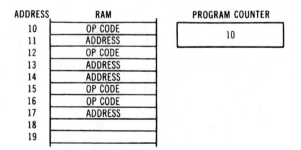

Fig. 4-11. **Program counter points to the address of the next instruction op code.**

(continued next page)

After the instruction in location 12 is executed, the program counter will contain the address:

    a. 12
    b. 14
    c. 15
    d. 18

---

13   (*c.* 15)   Go to Frame 14.

CPU Program Execution

---

14   Now let's illustrate how a typical CPU executes a simple program. To show this we introduce several common 8080/8085 instructions. These are:

- Move A to B (MOV B,A): This is a single-byte instruction that causes the content of the accumulator (A) to be moved to register B. The content of the accumulator is not erased.
- Load Accumulator (LDA): This is a three-byte instruction that takes the content of the address specified by the second and third bytes and loads it into the accumulator.
- Add Immediate (ADI): This is a two-byte instruction that takes the content of the second byte of the instruction and adds it to the content of the accumulator. The sum is stored in the accumulator replacing the number that was there previously.
- Halt (HLT): Halt is a single-byte instruction that stops processing.

Note that we refer to each instruction by a three-letter abbreviation called a *mnemonic*.

Figure 4-12 shows a complete CPU and the RAM containing a simple program. The first instruction in the program is the LDA stored in location 0000H.

To begin execution, the address of the first instruction is loaded into the register that points to the location of the instructions in RAM. This is the:

    a. program counter
    b. MAR
    c. accumulator
    d. register B

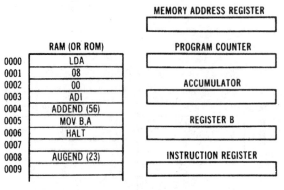

Fig. 4-12. CPU executing a simple program.

15 (*a.* program counter)   Next, the content of the PC is transferred to the MAR that addresses RAM. The LDA instruction op code is fetched and stored in the instruction register. The address bytes of the LDA instruction in locations 0001 and 0002 are then transferred to the memory address register. The address of the data to be loaded into the accumulator in bytes 2 and 3 is 0008H. The desired data word in location 0008H is retrieved and loaded into the accumulator. This is the augend (23).

After the LDA instruction is executed, the number 23 is stored in the accumulator. Next, the PC is incremented three times. The ADI instruction is then fetched and stored in the instruction register. The number to be added is stored in location 0004H.

This is an ADD immediate instruction where the second byte contains the data, in this case the addend (56). The number in location 0004H is 56, which is added to 23 in the accumulator, creating the sum 79. The mnemonic of the next instruction to be executed is MOV B,A in location 0005H.

The sum in the accumulator is then moved to the B register by the MOV B,A instruction. The content of the accumulator is not changed when data is transferred to the B register.

Finally, the HLT instruction is executed stopping the program. The content of the program counter at this time is:

a. 0007H
b. 0008H
c. 0009H
d. 0010H

16 (*a.* 0007H)   The program counter is incremented once since the HLT is a one-byte instruction.

As you can see by this step-by-step analysis of the execution of a computer program, there is nothing mysterious about its operation. The CPU simply fetches and then executes the sequentially stored instructions at very high speed until the operation is complete.

If you analyze what is really going on, you can see that most of the operations are data transfers, moving numbers, instructions, and other binary words from one place to another. The rest are arithmetic operations like addition. Together, all of these simple functions combine to implement the function specified by a single instruction. The instructions are then combined to form a program that does something useful, such as add two numbers.

Go to Frame 17.

Microcomputer Buses

17   Data transfers in a microprocessor take place in parallel. This means that all bits in a word are transferred simultaneously from one place to another. See Figure 4-13. It takes only a few nanoseconds (one billionth of a second) for all data bits in one register to be moved to another register.

The parallel data transfers take place over a data bus. A *bus* is simply multiple parallel electrical connections from a source to a destination. The bus may be multiple copper lines between ICs on a PC board or wires in a ribbon cable. Figure 4-14 shows a typical 8-bit data bus in a microprocessor. The eight parallel lines are usually represented by a single wide path as shown in Figure 4-14 to simplify the illustration. In some block diagrams, the bus is represented by a single line with a slash through it and an adjacent number indicating the number of lines in the bus.

A number of these buses are contained within the CPU and are known as internal buses. Refer to Figure 4-6. Locate the internal data bus. All inter-register data transfers take place over this bus.

In Figure 4-6, locate the address bus. How many bits does it carry?

    a. 8
    b. 16
    c. 24
    d. 32

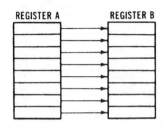

Fig. 4-13. A parallel data transfer from register A to register B.

Fig. 4-14. Data bus representation.

18   (*b.* 16)   These bits are labelled A_0–A_{15} in Figure 4-6 on page 56.

Microprocessors usually have three major buses, a data bus, an address bus, and a control bus. These are made available to external circuits. A typical 8-bit CPU has an 8-bit data bus and a 16-bit address bus as shown in Figure 4-15. The data bus sends data to and from the CPU, RAM, ROM, and I/O sections. All data transfers between the CPU and memory or I/O sections take place over the data bus. The address bus drives all of the memory and I/O devices. The control bus is a collection of signals into and out of the CPU that initiate and time the various memory accesses and I/O operations.

When an instruction is fetched from RAM or ROM, it is transferred over the data bus from the memory into the instruction register. Any data word retrieved from memory or a peripheral device via the input section also passes over the data bus into the accumulator or general-purpose register.

When a "store" instruction is executed, the word in the accumulator is transmitted over the data bus into RAM. Data can also be transferred from the accumulator over the

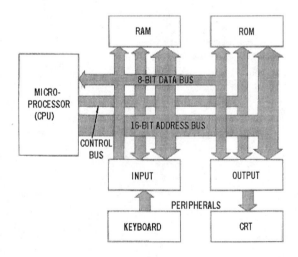

Fig. 4-15. Block diagram of a microcomputer showing data and address buses.

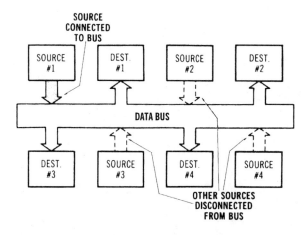

Fig. 4-16. Data source bus and destination connections.

data bus to a peripheral device such as a CRT (an output operation). Or, data from an input device such as a keyboard passes over the data bus and is placed into the accumulator. The important point here is that data can move in either direction over the data bus. We say that it is a *bidirectional* bus.

Another key point is that the data bus can be connected to only one data source at a time. Data can originate only at a single source, but it can be sent to one or more destinations. To accomplish this, circuits called bus multiplexers, or three-state line-driver circuits, are used to connect or disconnect the various data sources to or from the bus. Figure 4-16 illustrates this concept.

Since the bus in Figure 4-16 is bidirectional, any two sources may transmit simultaneously to any destinations.

   a. True
   b. False

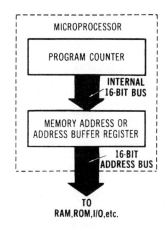

Fig. 4-17. Generation of the address bus in the 8080 microprocessor.

19   (*b.* False)   The address bus is a unidirectional bus. It transfers an address from the CPU to all external circuits (memory and I/O). Address words are produced in the CPU. The program counter generates the address that points to the instruction to be fetched. The content of the program counter (PC) is transferred over a parallel 16-bit internal address bus to the memory address register or address buffer. The output of the MAR or address buffer is the address bus. See Figure 4-17.

The MAR may also receive an input from the instruction word. The address bytes of an instruction word referencing the location of a data word in RAM or ROM are loaded into the MAR just prior to the instruction execution. Once the address bytes of the instruction are transferred to the MAR, they appear on the address bus and access the specific RAM or ROM location.

The control bus consists of numerous signals generated by the microprocessor and used to initiate various memory or I/O operations. The control bus also contains input lines from external circuits that tell the CPU what to do and when. The number of control signals varies from CPU to CPU.

Figure 4-15 shows a typical 8-bit microprocessor with its address, data, and control buses. Note that these buses are connected to both a RAM and a ROM. Input and output circuits are also shown. This arrangement is typical for virtually all microprocessor-based equipment.

Now answer the Self-Test Review Questions on the diskette before going on to the next unit.

Microprocessor Architecture

Stack and Stack Pointer

1   In this unit you will learn about microprocessor architecture. The architecture of the popular 8080/8085, Z-80, and 8051 microprocessors will be covered to give you a "real-world" look at how these devices work. Keep in mind that all other commercial microprocessors are similar, so all that you learn here applies to any CPU that you may encounter.

Architecture refers to three characteristics of a CPU:

1. the physical make up of the logic circuitry, particularly the type and number of registers
2. the instruction set, which contains the specific commands that the CPU executes
3. the addressing modes or how the CPU accesses instructions or data in memory

Let's look at the register make-up of the CPU first.

(continued next page)

You have already learned about the main registers in the CPU, such as the accumulator, instruction register, and program counter. But most CPUs also have several other important registers depending on their architecture. The *stack pointer* is an example. This is a special address register that defines a temporary storage area in RAM called the *stack*. It is an important feature of modern microprocessor architecture.

The *stack pointer* is a register similar to the program counter because it contains an address that points to some word in RAM (read/write memory). The location is one word of a memory area called the *stack*. The stack is simply a predesignated area of RAM set aside by the programmer for the temporary storage of data and instructions.

The stack pointer register is loaded with an address by a special instruction. The stack pointer is usually able to address any location in RAM. In most 8-bit microprocessors, the stack pointer register is 16 bits long.

The stack pointer register address identifies a(n):

    a. location in ROM
    b. location in RAM
    c. I/O port
    d. any of the above

---

2   (*b.* location in RAM)   Because any address can be loaded into the stack pointer, any location in RAM can be identified as the stack. This address defines the first location in the stack area or the "top of the stack." Sometimes the stack is assigned to the high end of RAM, which is that area with the highest address. For example, in a microcomputer with 4K (4096) of RAM, the stack would start at the top address 4095. Data is stored in the stack so that the first word goes in the highest address while successive words are stored in sequentially lower addresses. As the data is stored in the stack, the stack pointer is decremented to sequentially lower addresses. Figure 5-1 is a memory map identifying the stack. The size of the stack area is variable and depends on how it is used. Usually it is rarely ever more than several hundred bytes in size.

The "top of the stack" is the:

    a. highest memory address
    b. lowest memory address
    c. address in the stack pointer
    d. any address designated by the programmer

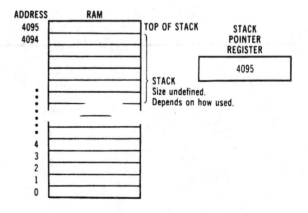

Fig. 5-1. Memory map identifying the stack.

---

3   (*c.* address in the stack pointer)   The stack is used to temporarily store addresses and data, such as the content of the program counter and the contents of the accumulator (or GPRs) if the normal sequence of program execution is interrupted. The CPU normally executes a main program whose instructions are stored sequentially, one after another in memory. But it is possible for this normal sequence to be changed. For example, the CPU could stop executing one program and jump to another new program and begin executing it. This new program is called a *subroutine.*

A *subroutine* is a relatively short program that performs some specific operation. It is usually stored in RAM or ROM outside of the normal program sequence. When the specific operation is needed, the CPU is directed to go outside the normal program execution sequence, execute the subroutine, and then return to the main program.

A subroutine is:

a. a part of the normal program sequence
b. data in ROM
c. the program addressed by the stack
d. a short sequence of instructions that do a specific operation

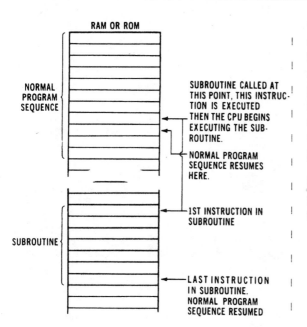

Fig. 5-2. The effect of a subroutine on program sequence.

In the figure:

RAM OR ROM

NORMAL PROGRAM SEQUENCE

SUBROUTINE CALLED AT THIS POINT, THIS INSTRUCTION IS EXECUTED THEN THE CPU BEGINS EXECUTING THE SUBROUTINE.

NORMAL PROGRAM SEQUENCE RESUMES HERE.

1ST INSTRUCTION IN SUBROUTINE

SUBROUTINE

LAST INSTRUCTION IN SUBROUTINE. NORMAL PROGRAM SEQUENCE RESUMED

4 (*d.* a short sequence of instructions that do a specific operation)   Figure 5-2 shows how the CPU instruction execution may progress assuming a subroutine is used. When the subroutine is called for, the instruction in progress is completed. Then the CPU jumps out of its normal sequential instruction execution sequence and begins executing the subroutine. When the subroutine has been completed, the CPU jumps back to the main program and picks up where it left off.

Assume that a 1-byte instruction in the main program at location 26 is executed. Then the CPU jumps to a subroutine. After the subroutine is completed, the next instruction to be executed is in location:

a. 25
b. 26
c. 27
d. 28

5 (*c.* 27)   If the CPU is to remember its place when a subroutine is called for, the program counter must be saved temporarily. Also, since the subroutine will use the accumulator and other CPU registers, their content should also be saved if intermediate results are to be remembered. The stack is used for that purpose. In most microprocessors, when a subroutine is called for, the content of the program counter is stored in the stack automatically. Then special instructions may be used to store other CPU registers in the stack, if desired.

(continued next page)

Once the subroutine is completed, the program counter content is retrieved from the stack automatically and restored so that the normal program sequence resumes.

Since the program counter is usually a 16-bit word, it will occupy two bytes in the stack. If the stack pointer is initially set to 1023, then the program counter content will be split in half and one byte will be stored in 1023 and the other in 1022.

If the stack pointer is initially set to 8191, then the program counter will be stored in stack locations:

>   a. 8191, 8192
>   b. 8190, 8189
>   c. 8192, 8193
>   d. 8191, 8190

---

6 (*d.* 8191, 8190) Remember that as data is stored on the stack, the stack pointer is decremented so that the first word is stored in 8191 and the second in the next lower address, or 8190.

Data is stored in the stack by special instructions designed for this purpose. Push instructions cause the contents of certain registers to be "pushed" onto or stored in the stack. Pull or pop instructions cause data to be retrieved from the stack and restored to the registers.

Each time a word is stored or pushed onto the stack, the stack pointer register is decremented. This causes the data from the registers to be stored in successively lower stack area locations. For example, one push instruction causes a word to be stored in the location designated by the stack pointer, say address 1022. The stack pointer is decremented to 1021 so that the next push instruction causes data to be stored in this location. If another push instruction is executed, the data will be stored in stack location 1020 and so on. The stack pointer always points to the last word that was stored.

When data is "pulled" or "popped" out of the stack, the stack pointer is incremented. The last word stored in the stack is retrieved first. As each pull or pop instruction is executed, the stack pointer is incremented, thereby retrieving data in the reverse order that it was stored. That is why a stack is often called a last-in, first-out (LIFO) memory.

Data stored in the stack first is:

>   a. retrieved last
>   b. retrieved first
>   c. retrieved when needed
>   d. never retrieved

---

7 (*a.* retrieved last) You are probably wondering at this point just how or when a subroutine is called for. There are two ways: (1) when a CALL instruction is executed or (2) when an interrupt occurs. The CALL instruction is used deliberately to cause the normal sequence of instruction execution to be changed so that a subroutine can be executed.

A subroutine is usually some function that may be repeated two or more times in a program. Rather than write the same set of instructions multiple times and waste memory space, it is more economical to write the subroutine once and then jump to it when needed. Today, most programs are written in modular form, whereby the program is divided up into logical sections with each section typically being a subroutine. The main program simply sequences the subroutines to do a particular operation.

A sequence of instructions is needed three times in a program. The sequence occupies 125 memory locations. How many locations are saved by using a subroutine?

    a. 125
    b. 250
    c. 375
    d. none

---

8   (*b.* 250)   An involuntary execution of a subroutine is caused by the occurrence of an interrupt. An interrupt is a microprocessor input signal that is usually generated by some external device such as a peripheral. The peripheral signals the CPU via an interrupt signal that it needs attention. The interrupt automatically causes the program counter to be stored in the stack and initiates a subroutine. Thus the peripheral device needing attention is serviced by initiating some input or output operation. The program previously in progress is then resumed.

There is also a software interrupt that permits a programmer to go to a subroutine in another way.

You will learn more about interrupts when the input/output section of microcomputers is discussed later.

An interrupt is a(n)

    a. input signal
    b. output signal
    c. signal received over the data bus
    d. signal transmitted over the data bus

---

9   (*a.* input signal)   Go to Frame 10.

Types of Instructions

10   Another inherent characteristic of microprocessor architecture is the type and number of instructions used by the CPU. An *instruction* is a binary word that tells the processor some specific operation to perform. All microprocessors have a wide variety of instruction types collectively referred to as the instruction set. There are four basic categories of instructions. They are:

    a. data movement
    b. data manipulation
    c. program manipulation
    d. program status and machine control

All microprocessors have instruction types in all four categories. These instructions are implemented in the wiring circuitry of the computer.

Data movement instructions cause binary words to be transferred from one place to another. These instructions may cause transfer of data between registers or between a register and some memory location. Input and output operations where data is transferred between peripherals and CPU are also initiated by data movement instructions. Instructions such as "load accumulator," store accumulator, and "move X to Y" are typical examples of data movement instructions.

There are five basic ways data is moved in a microcomputer. Which of these is used to store a word from the accumulator in RAM?

    a. register to register
    b. memory to register
    c. register to memory
    d. register to peripheral (output)
    e. peripheral to register (input)

---

11   (*c.* register to memory)   The following examples are some of the ways data can be transferred. These 8080/8085/Z-80 instructions are typical of those in any microprocessor.

Register to Register

MOV H,E: This single-byte instruction moves the content of the E register into the H register.

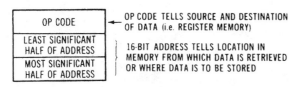

Fig. 5-3. Three-byte data transfer instruction.

Memory to Register

LHLD: Load the H and L register pair directly. This 3-byte instruction causes the 8-bit word stored at the location designated by the address in the second and third bytes of the instruction to be loaded into the L register. The word in the next successive address location is loaded into the H register. See Figure 5-3.

Register to Memory

STA: Store accumulator. This 3-byte instruction causes the content of the accumulator to be stored in the memory location designated by the address in the second and third bytes of the instruction. See Figure 5-3.

Examples of data movement instructions that produce input/output operations are as follows:

Register to Peripheral

OUT: Output. This 2-byte instruction causes a word in the accumulator to be transferred over the data bus to some external peripheral device. The second byte of the instruction is an address that is placed on the address bus and used to select one peripheral. See Figure 5-4A.

Peripheral to Register

IN: Input. This 2-byte instruction causes a byte from an external peripheral to be placed on the data bus and loaded into the accumulator. The second byte of the instruction contains a 1-byte address that identifies the input peripheral. See Figure 5-4B.

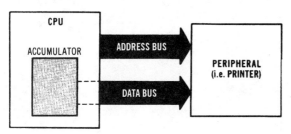

(A) Output operation.

(B) Input operation.

Fig. 5-4. Operations initiated by I/O instructions.

Which of the following is a type of data transfer instruction?

    a. load the stack pointer
    b. swap the contents of two registers
    c. push and pop instructions
    d. all of the above

12   (d. all of the above)   Data manipulation instructions do just what their name implies. They process data in a variety of ways. There are two basic types of data manipulation instructions: arithmetic and logic.

Arithmetic instructions cause various arithmetic operations to take place on the data in the microcomputer. The most common arithmetic instructions are add and subtract.

(continued next page)

These operations are performed by the ALU. The ALU can process two input words at a time. One of these words comes from the accumulator or a general-purpose register. The other word typically comes from some memory location. Some microprocessor architectures permit arithmetic operations to take place between two internal registers. In all cases, the result of the add or subtract operation is stored in one of the original operand registers.

Some typical 8080/8085/Z-80 add and subtract instructions are:

ADD r:	Add register. This single-byte instruction causes the content of register r (B, C, D, E, H, or L) to be added to the number in the accumulator. The sum appears in the accumulator.
SUI:	Subtract immediate. This is a two-byte immediate instruction. The data word in the second byte is subtracted from the number in the accumulator. The difference appears in the accumulator. See Figure 5-5.

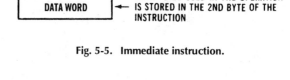

Fig. 5-5. Immediate instruction.

In an immediate instruction, the data to be used in the specified processing operation is stored in the second byte of the instruction.

The content of the accumulator is 17. An add immediate instruction (ADI) is executed. The second byte of the ADI is 8. The accumulator content after the add is:

    a. 8
    b. 9
    c. 17
    d. 25

---

13   (*d. 25*)   Increment and decrement instructions are also arithmetic operations. These instructions can cause one to be added to (increment) or subtracted from (decrement) a CPU register or a memory location.

Some typical 8080/8085/Z-80 instructions are:

INC r:	Increment register. This one-byte instruction increments register (A, B, C, D, E, H, or L) by one.
DCX rp:	Decrement register pair. This one-byte instruction causes one of the register pairs rp (BC, DE, or r HL) to be decremented by one.

If the content of register B is 1011000 and an INC B instruction is executed, the new content is 1011001.

Here's another example. If the register pair HL contains the binary equivalent of the decimal number 31,781 and two DCX HL instructions are executed, the new content of HL will be 31,779.

The simpler 8-bit microprocessors do not implement multiply, divide, or other higher-level math operations. However, most of the newer 16, 32, and 64-bit microprocessors also feature multiply and divide instructions as well as many other higher-level math functions such as square root, logarithm, and trig functions. Although complex internal circuitry is required to implement these instructions, they do greatly simplify and speed up mathematical operations in a microcomputer. In some 8-bit microprocessors, multiply and divide operations have to be programmed. Since a multiplication operation generally produces a product that is up to twice as long as the multiplier or the multiplicand, special double-word-length registers must be incorporated. (An 8-bit multiplier and an 8-bit multiplicand produce a 16-bit product.) In the same way, the dividend of a divide operation is typically twice as long as the divisor.

A short sequence of instructions used to multiply or divide any two numbers in a program would be called a(n):

- a. algorithm
- b. spreadsheet
- c. subroutine
- d. calculator

---

14 (c. subroutine)   Another class of data manipulation instructions performs logical operations such as AND, OR, NAND, NOR, complement, and compare. These logical instructions are typically used for bit manipulations. They allow the programmer to operate on the individual bits of a word rather than the whole word.

An AND instruction actually performs the logical AND function on corresponding bits in two 8-bit (or 16- or 32-bit) words. The AND instruction is typically used for masking operations where a portion of a data word can be eliminated by ANDing it with another binary word, called the mask word, with zeros in the appropriate bit locations. See Figure 5-6.

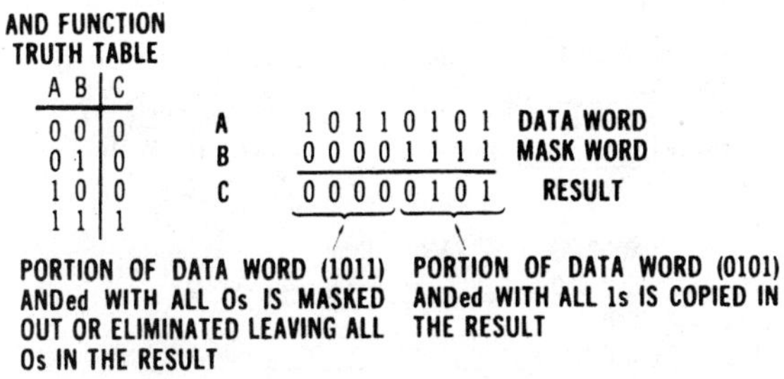

What is the result of ANDing these two words?

01011011 data word
01000100 mask word

- a. 01011111
- b. 10011111
- c. 00000111
- d. 01000000

AND FUNCTION TRUTH TABLE

A	B	C
0	0	0
0	1	0
1	0	0
1	1	1

A   1 0 1 1 0 1 0 1   DATA WORD
B   0 0 0 0 1 1 1 1   MASK WORD
C   0 0 0 0 0 1 0 1   RESULT

PORTION OF DATA WORD (1011) ANDed WITH ALL 0s IS MASKED OUT OR ELIMINATED LEAVING ALL 0s IN THE RESULT

PORTION OF DATA WORD (0101) ANDed WITH ALL 1s IS COPIED IN THE RESULT

Fig. 5-6. The logical AND function.

---

15 (d. 01000000)   All bits ANDed with a binary 1 are duplicated in the result. All bits ANDed with a 0 are replaced with a 0 in the result.

A typical AND instruction is given below:

ANI: AND immediate. The content of the accumulator is ANDed with the second byte of this two-byte instruction. The result appears in the accumulator.

(continued next page)

Another common logic operation is the OR function. It is sometimes used to merge two binary words into one. An example is given in Figure 5-7.

A typical OR instruction is given below:

ORI:	OR immediate. The binary word in the accumulator is ORed with the word in the second byte of this 2-byte instruction. The result appears in the accumulator.

The result of ORing 1001 with 0101 is:

    a. 0100
    b. 1101
    c. 1110
    d. 0001

OR FUNCTION TRUTH TABLE

D	E	F
0	0	0
0	1	1
1	0	1
1	1	1

```
D     1 0 1 1 0 0 1 1   DATA WORD
E     0 0 0 0 0 1 0 1   MERGE WORD
F     1 0 1 1 0 1 1 1   RESULT
```

PORTION OF DATA WORD ORed WITH PORTION OF MERGE WORD (ALL Os) CAUSES THAT PORTION OF DATA WORD TO BE REPEATED OR MERGED INTO THE RESULT

Fig. 5-7. The logical OR function.

16

(b. 1101) Now let's consider program-manipulation instructions. The main feature of this category of instruction is program sequence alteration. Instructions are normally executed one after another as they are stored in memory. When a program-manipulation instruction is encountered, this normal sequence is changed. These instructions are normally referred to as *jump* or *branch instructions.*

There are two types of jump instructions, *unconditional* and *conditional.* An *unconditional jump* changes the program execution sequence regardless of any other conditions existing in the CPU. When such an instruction is executed, the next instruction to be fetched is not the next one in sequence. Instead, the unconditional jump instruction specifies a new address where the next instruction to be executed is stored. Figure 5-8 illustrates this concept.

The other type of program-manipulation instructions is a *conditional jump* or *branch.* These are also known as decision-making instructions. These instructions allow the processor to "look at" the status of various conditions that exist in the CPU after a data transfer or arithmetic-logical operation. Then, based on the outcome, the decision-making instruction can cause the sequence of processing to be changed.

The typical jump or branch instruction consists of the op code byte and one or two additional bytes designating the address in memory to which the CPU will go for the next instruction.

If you want to jump to memory location 2A3E, what are the contents of the second and third bytes of the jump instruction?

    a. E3, A2
    b. A3, E3
    c. 3E, 2A
    d. 2A, 3E

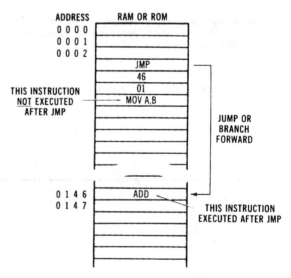

Fig. 5-8. The unconditional jump instruction.

17 (c. 3E, 2A) Typical decision-making instructions cause the normal instruction sequence to be altered like an unconditional jump, but only if some particular condition is met. These instructions can also be referred to as test instructions because they test the status or condition of the CPU. Some of these conditions are accumulator zero, accumulator not zero, accumulator content plus or minus, or the carry bit 0 or 1. Each of these conditions is indicated by the status of a flip-flop called a flag. Each time an instruction is executed, the flag flip-flops are automatically set or reset by the CPU to indicate the results of processing. For example, if the accumulator is cleared or set to zero, the Z flag is set indicating this condition. If a carry occurs out of the MSB position of the accumulator, the C flag is set. If the number in the accumulator becomes negative, the S or sign flag is set. In most CPUs there are five such flags. They are:

> Z: zero
> S: sign
> P: parity
> C: carry
> AC: auxiliary carry

Taken together, sometimes these flags are called the *status register* or the *condition code register.*

When a decision-making instruction is executed, it tests for the given condition. For example, it may test the flags to see if the content of the accumulator is zero. If the content is not zero, the next sequential instruction will be executed. That is, the instruction following the decision-making instruction will be executed. However, if the condition tested for does exist, in this case accumulator zero, then the normal instruction sequence is altered. The instruction following the conditional jump or branch will not be executed. Instead, the instruction located in an address designated by the instruction will be executed next.

A typical decision-making instruction is the JN or jump-if-negative. This instruction tests the sign of the number in the accumulator by looking at the S flag. If the S flag is 0, the word in the accumulator is positive. If S is 1, the number in the accumulator is negative. The JN instruction also contains an address where the execution sequence will resume if the condition is met.

Assume that a program contains an arithmetic instruction followed by the JN. The S flag is 0. Where is the next instruction to be executed stored?

    a. in the next sequential memory location
    b. in the location designated by the address in
       the JN instruction

(continued next page)

c. a memory location stored in the next instruction

d. the instruction register

---

18   (*a.* in the next sequential memory location)
Both the unconditional and conditional jump instructions can cause the sequence of operation to change to any location in memory. The jump can be forward in memory or backward. When a backwards jump (jumping to a lower address number) is executed, a program loop is formed. A program loop allows you to execute a particular sequence of instructions again and again if necessary. Each iteration or repetition of the loop may use different data and result in different outcomes. Loops are widely used in programs.

Figure 5-9 shows the operation of the jump-if-not-zero (JNZ) instruction. This instruction tests the zero (Z) flag in the status register. If the content of the accumulator is zero when the JNZ instruction is executed, no branch occurs. The condition "not zero" has not been met. Therefore, the normal program sequence is followed. The instruction following the JNZ is executed.

However, if the accumulator content is not zero when the JNZ instruction is executed, a branch occurs. The address bytes in the JNZ instruction point to the location in memory where the next instruction to be executed is stored.

Refer to Figure 5-9. If the accumulator content is zero, the address of the next instruction to be executed is:

    a. 0204
    b. 0002
    c. 0207
    d. 0003

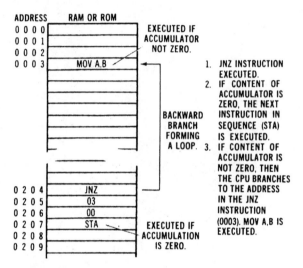

Fig. 5-9. The conditional jump-if-not-zero instruction illustrating a program loop.

---

19   (*c.* 0207)   If the accumulator content was something other than zero, then the JNZ would cause the CPU to loop back to location 0003 for the next instruction to be executed and the loop would repeat.

Two other useful program-manipulation instructions are CALL and return (RET). These instructions are also used to modify the usual program sequence. The call and return instructions are used with subroutines.

A *subroutine* is a sequence of instructions that performs some useful function. For example, it may be an operation that is needed repeatedly in another program. Some examples are I/O routines and math operations. Instead of writing this sequence of instructions many times, it can be written once and stored away in any desired section of RAM (or ROM). When the main program wishes to refer to the subroutine, a CALL instruction is issued. A CALL instruction is like an unconditional jump instruction in that the normal program sequence is changed. The CALL in-

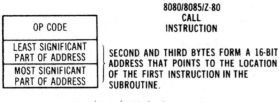

(A) 8080/8085/Z-80 CALL instruction.

(B) 8080/8085/Z-80 RET instruction.

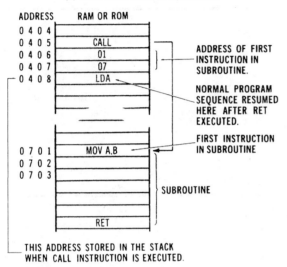

THIS ADDRESS STORED IN THE STACK WHEN CALL INSTRUCTION IS EXECUTED.

(C) How CALL and RET instructions are used in a subroutine.

Fig. 5-10. The CALL and RET instructions and how they are used in subroutines.

struction forces an address into the program counter that accesses the first instruction of the subroutine.

When a CALL instruction is executed, the content of the program counter is automatically stored away in the stack so that the normal program sequence can be resumed when the subroutine has been executed.

At the end of every subroutine, a return (RET) instruction is executed. The return instruction causes the program counter content to be retrieved from the stack and placed in the program counter. The normal program sequence is then resumed. Figure 5-10 illustrates this concept.

Every subroutine must end with this instruction:

    a. CALL
    b. RET
    c. JMP
    d. POP

20   *(b.* RET)   The final category of microprocessor instructions includes the program status and machine control instructions. Machine control instructions are those that affect the overall operation of the CPU. Two of the most often used machine control instructions are the HALT and NOP (no operation). The HALT instruction is used to stop the processing. When a program is complete and no further operation is desired, a HALT instruction can be included as the last instruction in the program.

A NOP instruction does nothing except to use up time. No processing occurs, but a NOP does cause the program counter to be incremented. NOP instructions are usually used to aid in timing operations and can be used to replace redundant, unnecessary, or incorrect instructions in a program without affecting the locations of other instructions in the program. They simplify the debugging and revision of programs.

(continued next page)

Two typical program status instructions are the enable interrupt (EI) and disable interrupt (DI) instructions. These one-byte instructions do exactly as their names imply. Interrupts from external devices are either recognized (EI) or masked (DI).

An interrupt from a peripheral will not be recognized if which instruction is executed?

     a. HALT
     b. NOP
     c. EI
     d. DI

---

21   (*d. DI*)   This concludes our discussion of instruction types.

Now go to Frame 22.

Addressing Modes

---

22   Many instructions executed by a microprocessor refer to a data word to be used in the operation specified by that instruction. The data word is stored either in some memory location or in a register. Some means must be used to refer to or locate this data word. This is usually done with an address associated with the instruction itself.

Remember that most 8-bit microprocessors have three instruction-word formats. See Figure 5-11. The first 8-bit byte in an instruction is the op code, which is the binary bit pattern that tells the CPU what to do. Any additional bytes in the instruction are normally reserved for addresses that reference the desired data word, or operand. The *operand* is a term describing one of the data words to be processed in a problem. There are many different ways of addressing information in a microcomputer. Those addressing modes common to the typical 8-bit microprocessors are discussed here.

Single-byte instructions normally contain only an op code. While no address is used, these instructions can also refer to data words. They do this in an implied manner. We say that such instructions use the *implied addressing mode.*

The instruction op code itself typically designates the location of the data word to be used. The most common storage place for such data words is in the accumulator or one of the other general-purpose registers. An example of such an instruction is one that causes the addition of two general-purpose registers. The two registers involved in the operation are designated by the op code. Because identifiable

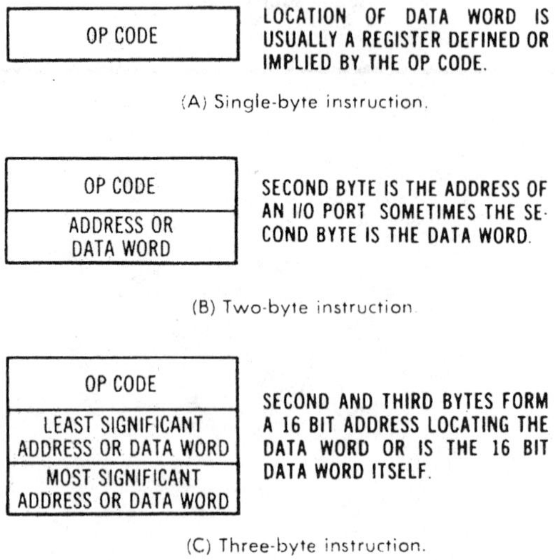

OP CODE

LOCATION OF DATA WORD IS USUALLY A REGISTER DEFINED OR IMPLIED BY THE OP CODE.

(A) Single-byte instruction.

OP CODE
ADDRESS OR DATA WORD

SECOND BYTE IS THE ADDRESS OF AN I/O PORT SOMETIMES THE SECOND BYTE IS THE DATA WORD.

(B) Two-byte instruction.

OP CODE
LEAST SIGNIFICANT ADDRESS OR DATA WORD
MOST SIGNIFICANT ADDRESS OR DATA WORD

SECOND AND THIRD BYTES FORM A 16 BIT ADDRESS LOCATING THE DATA WORD OR IS THE 16 BIT DATA WORD ITSELF.

(C) Three-byte instruction.

Fig. 5-11. Instruction-word formats.

registers are involved, this addressing mode is often called the *register addressing mode.*

Another commonly used addressing mode is called the *immediate mode.* Immediate instructions normally consist of two bytes, the op code and a second byte that contains the data word itself. Here the data to be used in the desired operation is contained in the second byte of the instruction itself. The immediate addressing mode executes quickly because no special addressing operation is required. The data is simply the next byte in sequence after the op code. When 16-bit operations are to be performed, the immediate instruction contains three bytes. The first byte is the op code as usual, while the next two bytes contain the 16-bit word divided into two 8-bit parts.

An immediate data word is how long?

    a. 8-bits
    b. 16-bits
    c. 8- or 16-bits
    d. It is determined by the program user.

23

(*c.* 8- or 16-bits)   The most commonly used addressing mode is the direct mode. With direct addressing, the second or the second and third bytes of the instruction form an address that references the memory location where the desired data word is stored. See Figure 5-11. When an instruction using direct addressing is executed, the address word stored in the second and third bytes is loaded into the memory address register. The desired data word is retrieved from memory and used in the operation performed by the instruction. With a 16-bit address, the total addressing capability is 16 bits or 2^{16} (65,536) words of RAM and/or ROM.

In some microprocessors, the direct addressing mode uses only one address byte. Such 2-byte instructions can only access data within a $2^8 = 256$ word range in RAM or ROM. Although that may seem like a restricted size, in most embedded controllers 256 bytes is far more RAM than usually needed to implement a useable program.

The contents of the second and third bytes of an add instruction are 0F and C8 respectively. Therefore, the number to be added is stored in which memory location?

    a. 0FC8
    b. 8CF0
    c. F08C
    d. C80F

24

(*d.* C80F)   Another commonly used addressing mode is referred to as *register indirect.* In this mode, 1-byte instructions reference a special register in the CPU

(continued next page)

that contains the address of the data. When such an instruction is executed, the address stored in the special register is transferred to the memory address register and the address bus to retrieve the desired data word from memory. Of course, to use this address mode, the address register must be loaded with the desired address prior to executing the instruction. The loading of this register takes place by executing an appropriate data transfer instruction. In the 8080/8085/Z-80 CPUs, the BC, DE, and HL register pairs are used for this purpose. For instance, the instruction that puts an address in the HL register is the LXI H, L.

One of the most sophisticated addressing modes used in microprocessors is the indexed mode of addressing, also called *indexing.* Like the register indirect mode, the index method requires a special internal register. This register is referred to as the *index register.* In most microprocessors, this is a 16-bit register that can be loaded by a special index register instruction. Another instruction can cause the content of the index register to be transferred to a desired memory location. Increment and decrement instructions are also available to modify the index register.

The format of an instruction using the indexed mode of addressing is similar to that with direct addressing in that the second and third bytes contain an address. This address does not directly address the desired data word. Instead, the address contained within the instruction is typically added to (or subtracted from) the contents of the index register to form the address where the desired data word is stored. This process is called indexing. See Figure 5-12.

The 8080/8085 and 8051 microprocessors do not have an index register, but the Z-80 has two of them. Most modern microprocessors contain at least one index register and many have special-purpose address registers that can be used as index registers.

The indexed mode of addressing is an extremely efficient way to access data in memory. It is particularly useful in retrieving or manipulating long strings of data stored in sequential memory locations. Lists, tables, or arrays of numbers stored in memory can be quickly processed or moved from one place to another by using the index mode of addressing.

To form the address used to reference the desired data word in the indexing mode, the number in the index register is added to the:

    a. address in the instruction word
    b. content of the program counter
    c. content of the accumulator
    d. word in the MAR

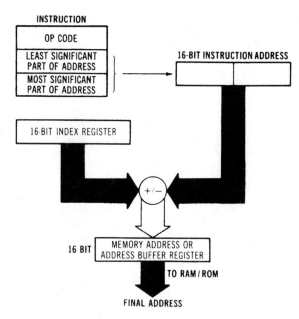

Fig. 5-12. **Index addressing mode.**

25   (*a.* address in the instruction word)   Some microprocessors also use a special addressing mode known as the relative addressing mode. The relative addressing mode is implemented with a 2-byte instruction. The first

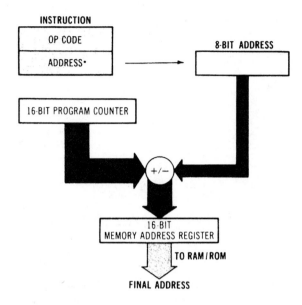

Fig. 5-13. Relative addressing mode.

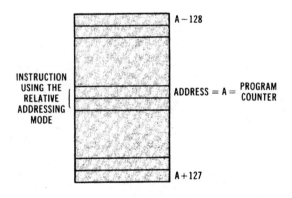

Fig. 5-14. Address range of the relative addressing mode.

byte is the op code as usual. The second byte contains a number that indicates the relative position of the data word to be retrieved. The content of the second byte in the instruction is usually added to the content of the program counter with the result being fed to the memory address register to access the desired data word. Refer to Figure 5-13.

Keep in mind that the program counter contains the address of the next instruction to be executed. Once the instruction is executed, the program counter is incremented so that it points to the next instruction in sequence. The relative addressing mode indicates that the data word to be used is stored in a memory location relative to the address in the program counter. The data address is relative to the instruction address.

The number stored in the second byte of the instruction contains eight bits. The first or most significant bit is used as the sign bit. This means that both positive and negative numbers in the 2's complement format can be added to the program counter. This gives the relative addressing mode the ability to reference a data word before or after the instruction address. If the number in the second byte of the instruction is positive, the address formed by adding it to the program counter will be greater than that of the address of the instruction word. However, if the second byte contains a negative number, an address lower than the instruction address is created when the second byte is added to the program counter. (Adding a negative number is the same as subtracting.)

Because the second byte is an 8-bit word, the total addressing range of the relative mode is 256 bytes. But since one bit is used as the sign bit, this leaves seven bits to designate the addressing range. This gives us a range of 128 memory locations centered on the address in the program counter. See Figure 5-14. The relative addressing mode is usually used with jump instructions.

There are other special addressing modes that are unique to some types of microprocessors. When learning a new CPU, always review the addressing modes to see what is typical and common and what is different and special.

Now answer the Self-Test Review Questions on the diskette before going to the next unit.

Input/Output Operations

LEARNING OBJECTIVES

When you complete this unit, you will be able to:

1. Define the terms *I/O port, interface, controller, serial, parallel, baud,* and *baud rate.*

2. Explain the operation of each of the four major I/O operations used with microcomputers: programmed I/O, polled I/O, interrupt I/O, and direct memory access.

3. Explain the basic characteristics and operation of the Centronics printer and the Small Systems Computer Interface (SCSI) parallel interfaces.

4. Explain the basic characteristics and operation of the EIA RS232, Universal Serial Bus (USB), and the IEEE 1394 serial interfaces.

5. Explain the operation of a UART.

Basic I/O Operations

---

1   Input/output (I/O) operations refer to the means by which the CPU and memory communicate with the outside world. All microcomputers must be connected or interfaced to some external devices in order for them to accomplish useful work. Some examples are:

a. Connecting a personal computer to external peripheral units such as a keyboard, mouse, video monitor, disk drives, or a printer in order to communicate with a human operator.

b. Connecting an embedded microcomputer to external components such as temperature or pressure sensors or switches whose output sig-

(continued next page)

nals will be monitored, or to output actuators such as relays, solenoids, valves, motors, and indicator lights whose condition will be controlled.

Regardless of the exact application, the input/output operations take place in the I/O section of the computer. This section of the computer is made up of individual circuits that connect the CPU to a specific external device. These circuits are referred to as *interfaces*. You will also hear the interface referred to as an *I/O port,* or, in some cases, a *controller* or *device controller.* In a personal computer, some of the interfaces are packaged with the CPU and memory. Others are add-on circuits that plug into the computer bus. These are often referred to as *adapters.* In an embedded microcontroller, the I/O circuits are usually inside the chip.

The interface is composed of various digital logic circuits that make the CPU compatible with the external equipment regardless of its configuration. The interface is connected to the microcomputer via the data and address buses as shown in Figure 6-1.

There are four basic ways that I/O operations are performed: programmed I/O, polled I/O, interrupt I/O, and direct memory access (DMA). Let's look at each of these in more detail. Go to Frame 2.

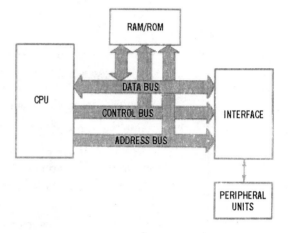

Fig. 6-1. The I/O interface that connects the computer to external devices.

Programmed I/O Operations

2 The simplest method of input/output is the programmed I/O method. In programmed I/O operations, the CPU is in direct control of all data transfers and other I/O functions. The programmed I/O method involves using microcomputer instructions to cause data transfers and control operations. If an output operation is desired, the data to be outputted is simply transmitted from the CPU to the external equipment using an output instruction. If an input operation is required, an input instruction is issued and the data is transferred one word at a time from the external device to one of the CPU registers and then stored in memory. The input or output instructions are simply included in the program at the points where you wish to read in data or to send out data.

The CPU can also send control signals to the peripheral device through the I/O port. The control signals are usually binary words called *control codes.* Each code causes the peripheral device to do something other than receive data.

Although programmed I/O operations are simple, timing or speed variations between the CPU and the external peripheral devices must be resolved. The CPU runs at very high speeds while peripheral devices are slow, usually due to their mechanical nature. When an output instruction is executed, one word of data will be transmitted from the accumulator or a general-purpose register over the data bus to the peripheral device through the interface. The equipment, of course, must be ready to receive the data at that time or the data will be lost. In the same way, an input operation will transfer one word on the data bus into a CPU register when an input instruction is executed. Again, the peripheral device must be ready to send the information. If it is not, the instruction will be executed and the desired data may not be input. These timing variations are generally resolved by the logic circuitry in the interface or by special programming considerations.

Speed differences between the very high-speed CPU and the usually much slower peripheral device are often difficult to resolve, not only because of the wide speed differences, but also because of the random need for a data transfer. For example, if a human operator is going to enter data into the computer via a keyboard, the computer has no idea when the operator may begin or how fast the operations will take place. The computer cannot anticipate when a user might press a key on a keyboard or move a mouse. The speed of data entry will also vary considerably from one operator to another, depending upon his or her typing skills or general manual dexterity. The microcomputer has somewhat more control of output operations because the operator can often tell the computer when to begin outputting data. However, any speed differences must be resolved so that data outputted at one speed is not lost by transferring it to a lower-speed peripheral device.

Programmed I/O operations are simple and useful, but one common problem is:

  a. Excessive memory is required.
  b. The programming is complex.
  c. Most microprocessors do not support I/O.
  d. Timing incompatibilities complicate the process.

---

3 (*d.* Timing incompatibilities complicate the process.) Because of the speed differences and random nature of I/O data requests, they are referred to as *asynchronous.* Asynchronous data requests occur when they naturally happen. Asynchronous simply means there is no fixed time of occurrence or repeatable cycle.

One method of resolving the time difference between the CPU and the peripheral device is to use logic buffering in the interface. This buffering may take the form of a temporary storage register or even a small RAM where one or more words of data are temporarily stored as the data is transferred into or out of the computer.

A common method of resolving the timing differences in an asynchronous I/O situation is to take care of them when

(continued next page)

programming the I/O function. Most CPUs can be programmed to include timing delays or testing functions to resolve speed differences. For example, the microcomputer can be programmed to wait for an input to occur. The program may be one where the microcomputer is repeatedly testing the peripheral device to determine whether it is ready to send data. As soon as the microcomputer recognizes that a peripheral is ready, it will execute the input operation. For an output operation, the computer may again be programmed to test whether the peripheral device is ready to receive the data. Here, the computer is programmed to wait until that time occurs and the data is outputted to the device.

Asynchronous data requests and their attendant timing problems are resolved in most microcomputers by:

   a. using high-speed logic
   b. using low-speed logic
   c. making the peripherals faster
   d. having the CPU wait

---

4   (*d.* having the CPU wait)   This waiting is accomplished in software by having instructions test the peripheral and withhold a data transfer until the peripheral is ready to receive data. The process of making the computer wait for the peripheral device is simple to implement and quite effective. It typically uses a program loop where a simple testing program is repeatedly executed. This program loop consists of one or more instructions that test the status of the peripheral device. Usually some electrical signal is available from the peripheral device that can be used to signal its status to the microcomputer. This signal is often a "ready" or a "busy" signal that tells the computer when the peripheral is capable of sending or receiving data. This ready signal is read into the CPU via an input instruction. It is then tested, perhaps with one of the logical instructions, to determine if it is a binary 1 or binary 0. Typically, if a binary 0 is detected, the peripheral device is not ready. When a binary 1 is indicated, the peripheral device is ready. A branch instruction usually makes the test and the "ready" or "not ready" decision. If a "not ready" condition is detected, the jump instruction will create a program loop and the same sequence of instructions will be repeated. When the peripheral device is ready, a branch instruction will change the program sequence and cause the desired input or output operation to take place.

This process is illustrated in Figure 6-2. A single bit BUSY signal is sent to the interface by the peripheral. It is connected to one of the flip-flops in an 8-bit register in the interface. The CPU performs an input operation that reads the contents of register #1. The BUSY bit is tested to see if it is 0 or 1. If it is 0 or not BUSY, the CPU performs an output operation that transfers data via the 8-bit data bus through register #2 in the interface to the peripheral. Regis-

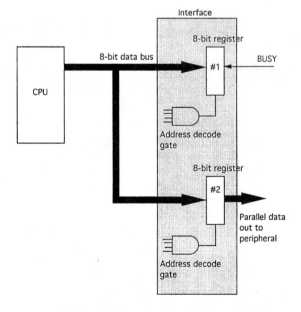

Fig. 6-2. Block diagram of a typical interface circuit showing control and data transfer I/O paths.

ter #1 is an input port while register #2 is an output port. Each port has an AND gate used for decoding the I/O address from the CPU address bus. When the AND gate recognizes its unique address, the port is enabled.

The primary disadvantage of the programmed I/O technique is that it does not make efficient use of the microprocessor. While the processor is waiting on the peripheral device, it cannot be executing other programs. Because of the vast speed differences between the microprocessor and most peripheral devices, many other programs could be executed during the waiting time. Although the waiting period may only be several milliseconds, this time translates to many hundreds or thousands of instructions for a CPU that executes instructions in microseconds or nanoseconds. Many complete programs could be executed during this time. Although this method of input/output is inefficient, it is simple to implement and quite effective. In a dedicated microcontroller, the inefficiency may not matter. In a personal computer, such inefficiency could not be tolerated.

The programmed I/O method is still one of the most widely used I/O methods despite its inefficiency. The inefficiency is often irrelevant in dedicated controllers since the microcomputer is usually partially dedicated to performing input/output operations. The CPU may not have other programs to execute until an I/O operation occurs.

The main disadvantage of programmed I/O is:

a. too slow
b. inefficient use of CPU
c. limited to one peripheral device
d. one-way data transfer only

---

5 (b. inefficient use of CPU)   Go to Frame 6.

Polled I/O Operations

---

6   Polled I/O is simply a variation of the programmed I/O method. It is used when more than one peripheral device is used. Most microcomputers will have two or more external devices connected to them. The CPU must service them all as input and output operations are required.

When more than one peripheral device is connected to the CPU, the I/O programs are set up so that each peripheral is "polled" to see if it is ready. The CPU sequentially looks at or polls each external device looking for the input signal that tells the program to initiate an input or output

(continued next page)

operation. Typically, each peripheral device will generate some kind of binary signal that will be sent to the interface that tells the computer it needs service. It may have data to send or it may be requesting that data be sent to it. This signal will usually set or reset a flip-flop in the interface that the CPU will monitor by reading it in and examining its state, which will indicate if and when to send or receive data. The main control program in the computer is set up to continuously circulate and poll each device looking for the need to do I/O.

Polled I/O works fine, but again, it is very inefficient. It wastes a great deal of valuable CPU time. To overcome this problem, interrupt I/O is used.

Polled I/O is used for

    a. inputs only
    b. outputs only
    c. when more than one I/O device is to be used
    d. very high speed data transfers

---

7   (*c.* when more than one I/O device is to be used)
Go to Frame 8.

Interrupt I/O Operations

---

8   The inefficiency of the programmed or polled methods of I/O is not a disadvantage in many dedicated embedded controllers. If the CPU doesn't have anything else to do, why not use it to test or poll the peripheral status? But this inefficiency is a major factor in a general-purpose computer that must work with a variety of peripherals. Further, in systems with multiple peripheral devices, the polling process may be too slow. High-speed I/O devices require a fast reaction from the CPU so that data is not lost or to ensure that a critical control function is performed in time. Polled, programmed I/O systems may simply not be fast enough when fast, multiple high-speed external devices are constantly competing with one another for CPU attention.

This inefficiency that results in slow CPU reaction time can be overcome by using a third method of input/output, called the interrupt I/O method. With this method, data transfers to and from the CPU still take place over the data bus through an interface, as with the programmed I/O method. The same CPU input/output instructions are used. The primary difference is that an interrupt is used to signal the CPU when an external device is ready to send or receive data. The interrupt is an input signal to the CPU. It is usually generated by the external device. When the external device requires data or has data ready to send to the CPU, the interrupt is produced. The interrupt does what its name implies—it interrupts the CPU. The CPU can be executing any other program instead of waiting.

All microprocessors are designed to accept interrupts and deal with them accordingly. Typically the microprocessor will complete the execution of any instruction currently in process when an interrupt occurs. It will then suspend further operations on the regular program sequence and alter its operation to service the interrupt. The interrupt forces the CPU to branch to a special subroutine stored in memory that is designed to perform the desired operation. Once that subroutine has been executed, the normal program execution sequence resumes.

When an interrupt occurs, the CPU:

    a. keeps processing
    b. halts
    c. branches to a subroutine
    d. waits

---

9   *(c. branches to a subroutine)* When an interrupt occurs during the execution of an instruction, no immediate action is taken while the instruction is allowed to finish. Then the program counter is incremented so that it points to the next instruction in sequence. At this time, the content of the program counter is automatically stored in the stack so that the CPU will remember its place. The CPU then automatically jumps to a memory location predesignated to contain the subroutine that will perform the desired I/O operation. This branching to the subroutine is roughly equivalent to executing a CALL instruction, except that it is done automatically by the CPU when an interrupt is received. The interrupt service subroutine ends in a return (RET) instruction that automatically causes the content of the stack to be retrieved and placed back into the program counter. The address placed there is, of course, the address of the next instruction in sequence in the previously interrupted program. Normal processing then resumes until another interrupt occurs.

The receipt of an interrupt causes the CPU to perform as if which instruction were executed?

    a. CALL
    b. RET
    c. HALT
    d. BRANCH

---

10   *(a. CALL)* In addition to storing the content of the program counter in the stack when an interrupt occurs, you may need to store the contents of other registers so that

(continued next page)

important data and valuable intermediate information is not lost when the I/O routine is executed. The content of the accumulator or some of the general-purpose registers as well as the content of the flag register usually will also be placed in the stack so the status of the machine is not lost. In some microprocessors, all of this is done automatically. In other microprocessors, only the content of the program counter is stored automatically. Special PUSH instructions are used to store the contents of other CPU registers in the stack. Then PULL or POP instructions are used at the end of the interrupt service subroutine to retrieve this information and return the status of the CPU to its former condition.

Most CPUs are also capable of recognizing and dealing with multiple interrupts. The CPU may be operating with several peripheral devices that will require attention at various times. Each peripheral device can generate an interrupt. When an interrupt occurs, the CPU must determine which peripheral is requesting service. The CPU will then select or be "vectored" to the appropriate subroutine service program. Such interrupts are called *vectored interrupts.*

In some complex applications, interrupts can also interrupt one another. It is possible for an interrupt from one peripheral device to occur during the time that the interrupt subroutine for another external device is being executed. These are referred to as *multiple-level interrupts.* Typically, such applications involve real-time operations. Here the processor is used in an application where the occurrence of an interrupt is immediate and cannot be ignored. Some action must absolutely take place to avoid losing important data or to initiate some critical operation.

When interrupts interrupt one another in real-time applications, the programmer must decide which peripheral is the most important. Since the computer can execute only one I/O subroutine at a time, it is important that I/O priorities be established. In this way, the CPU can give its attention to the most important external device.

Most of the larger CPUs used in general-purpose computers have a system for prioritizing the interrupts. The peripherals and I/O devices are connected to the interrupts in priority order so that they are given the service in order of their need.

Finally, in most microprocessors, interrupts must be enabled before they can be recognized. Special enable (arming) or disable (masking) instructions are used for this purpose. Some micros have at least one nonmaskable interrupt input for critical applications.

Which of the following is *not* true?

    a. Interrupts are widely used in real-time applications.
    b. A vectored interrupt is only one of several that a CPU can service.
    c. Multiple interrupts have random priorities.
    d. One interrupt may interrupt another.

11 (*c*. Multiple interrupts have random priorities.)
Go to Frame 12.

Direct Memory Access

12 One of the primary limitations of the programmed and interrupt methods of input/output is speed. In both cases, the speed of the data transfer is tied to the speed of the microprocessor and the I/O program. Data words can be transferred to or from the CPU as fast as the external devices can send or receive them. In any case, the maximum speed is set by the CPU execution speed of the input and output instructions. In some very critical high-speed applications, even the microsecond transfer rate of typical microprocessor I/O instructions is insufficient to deal with the situation. In these cases, a fourth method of input/output can be used. This is called direct memory access (DMA).

Refer to Figure 6-3. In the direct memory access method of I/O, the CPU is bypassed. A special DMA controller interface is connected between the memory and the high-speed external peripheral. Typical high-speed external devices are magnetic disk drives, video, CD-ROM drives, and analog-to-digital and digital-to-analog converters. The speed of most semiconductor memories enables them to more than adequately store or retrieve data with sufficient speed to satisfy virtually any high-speed external device. The special DMA controller interface, usually a single LSI IC, then causes data transfers to take place directly between the memory and the I/O device. In an output operation, data is transferred from RAM directly over the data bus to the external device via the high-speed controller IC. Alternately, in an input operation, data from the external device is passed through the controller interface and stored in RAM. The DMA controller replaces the CPU and the I/O program.

A common DMA peripheral application is:

a. a printer
b. a keyboard
c. a serial data transfer
d. a disk drive

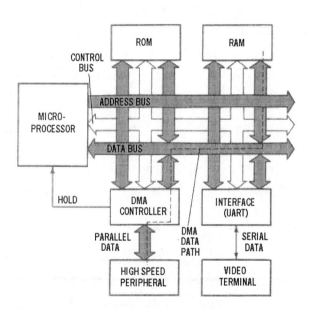

Fig. 6-3. Microcomputer system showing DMA interface to a serial video terminal and a high-speed peripheral device like a disk drive.

13 (*d*. a disk drive) The DMA controller interface contains a variety of counters and registers that are used to select and control the address of RAM. When data is to be

(continued next page)

transferred between the peripheral device and the memory, the proper starting and ending addresses of the memory locations involved must be known. In a high-speed data output operation, the starting address of the data in memory is given to the DMA controller. This is usually stored in a counter that is then incremented each time a data transfer occurs.

When the ending address of the data is reached, the output operation ceases. A similar situation exists for input operations. A block of memory is usually set aside to store the data. This memory block is designated by starting and ending addresses. These starting and ending addresses are typically supplied by a program executed by the CPU prior to the DMA operation. The CPU is normally used to set up the DMA controller interface in preparation for an I/O operation.

The DMA I/O operations also use interrupts. The high-speed peripheral device signals the DMA controller with an interrupt that a data transfer is necessary. A HOLD signal then tells the CPU to complete executing the present instruction and to go into a dormant state. At this time, the CPU relinquishes control of both the data and address buses. Then the DMA controller interface takes over and provides the addresses to RAM. Data transfers then take place between the peripheral device and the memory over the data bus via the interface. Once the data transfers are complete, control is returned to the CPU. The setup of the DMA controller is handled by a special I/O subroutine that supplies starting and ending addresses, word counts, and other information. In a DMA operation, normal program I/O instructions are used to set up the DMA controller. Then an interrupt is used to signal a request for an I/O operation.

Which of the following is true?

    a. DMA transfers pass through the CPU.
    b. DMA transfers take place between the peripheral device and RAM.
    c. DMA I/O is the slowest form of I/O.
    d. DMA I/O cannot use interrupts.

---

14   (*b.* DMA transfers take place between the peripheral device and RAM.)   Go to Frame 15.

Addressing External Devices

---

15   Most microcomputers have two or more I/O devices. At least one I/O port is needed to get data into the computer, and at least one port is needed to get the data out. Typical computers have many I/O ports and peripherals. For this reason, the computer must have some way of identifying the peripheral or I/O port to be used in an I/O operation. This calls for some kind of addressing scheme. You can identify and enable any one of several peripheral devices just as you can locate any word in RAM or ROM.

Some microprocessors have a built-in addressing scheme associated with the I/O instructions. The op code is the first byte of the instruction while the second byte is an address that states some specific external device to be used in the data transfer. This is how the I/O instructions in the 8080/8085/Z80 microprocessors work.

Assume an input operation is to be performed. When an input instruction is executed, the address byte of the instruction is placed on the address bus. A decoder in each peripheral interface looks at the address. The decoder in the addressed peripheral recognizes the address and enables the interface circuitry. This was shown in Figure 6-2. The data from the peripheral is placed on the data bus and read into the accumulator register. From there it is dealt with by the remainder of the program.

An output instruction works in a similar way. The address in the second byte of the instruction is placed on the address bus and sent to all peripherals. One of the peripherals recognizes the address and is enabled. A data word previously placed in the accumulator is transmitted over the data bus to the interface of the peripheral device.

How many peripheral devices can a microcomputer with an 8-bit I/O address identify?

    a. 8
    b. 16
    c. 64
    d. 256

---

16   (*d.* 256)   A more common way of addressing I/O ports is to use what is commonly called memory-mapped I/O. In this system, each peripheral interface is given an address that is commonly within the address space of the microprocessor. The address space is the total amount of memory that a microprocessor can address, which may be either RAM or ROM. With memory-mapped I/O, each I/O port is assigned to one of the unused memory addresses.

In most microcomputers, not all of the memory space is used. The program may take up some of it in ROM, and there will be an area allocated to RAM. Much of the remaining space is simply not needed. In these cases, some of the unused memory addresses are assigned to peripheral devices.

A key benefit of memory-mapped I/O is that no special I/O instructions are needed. Any instruction that can ac-

(continued next page)

cess memory can be used in I/O operations. Accumulator load and store instructions can be used for input and output operations respectively. Move instructions that access memory can be used to get data into or out of the CPU. Even arithmetic and logic instructions can be used. All of the various addressing schemes can be brought to bear on I/O operations if they can be useful.

An I/O operation is similar to that described earlier. To do an output operation, you might use a MOVE accumulator-to-memory instruction. The address that is part of that instruction is placed on the address bus monitored by the RAM and ROM and all peripheral devices. The peripheral recognizing the address is enabled and it receives the data in the accumulator over the data bus.

You could even do arithmetic and logic operations with input or output data. For instance, you may execute an ADD instruction with an address that selects one of the I/O ports. The address goes out on the address bus and enables the desired peripheral device. It then transmits a word of data over the data bus, which is then added to the number already in the accumulator.

No special I/O instructions are needed for memory-mapped I/O.

    a. True

    b. False

---

17   (*a.* True)   Most modern microcomputers use the memory mapped I/O method. Go to Frame 18.

Interface Fundamentals

---

18   Data transfers between an external peripheral device and a computer normally take place through the CPU. The only exception to this is when DMA is used, in which case the data is transferred directly between the peripheral device and RAM. Data transfers take place over the CPU data bus. However, the microcomputer data bus is typically isolated from the external equipment by the input/output section, which is generally referred to as an interface, controller, or I/O port. The primary purpose of the interface is to provide a form of buffering between the CPU and the external equipment. Buffering refers to making the microcomputer data bus compatible with the external equipment. This involves the formatting, timing, and logic level conversion of the data.

Figure 6-2 shows how the interface is positioned in a system. Physically, the interface could be a separate plug-in printed circuit board containing chips that perform all of the various buffering duties. Interfaces like this are common in personal computers. Typical examples are the disk controllers and the video display controllers that operate the CRT (cathode ray tube). The interface may also be just a separate set of I/O circuits in an embedded single-chip microcontroller.

Regardless of the physical packaging, most interfaces do essentially the same thing. They facilitate the transmission of data into or out of the computer. There are probably as many unique ways to do this as there are ways to select lottery numbers. However, over the years a variety of standard interfaces have been developed. Standards arise when the interface developed by one manufacturer is adopted by everyone else for the same function. You will hear this referred to as a de facto standard. Also, standards arise when a group of individuals from companies with the same needs gets together and decides upon a common interface that everyone will use. Such standardization is desirable to the user because many different devices can be made to use one type of interface. This makes peripheral devices compatible with many different computers as long as they have the standard interface. If the standard is widely adopted, then more computers can be made compatible with more peripheral devices. Although there are literally dozens of different standard interfaces and new ones being introduced every year, only a few are very widely used and generally considered to be "popular." A later section in this unit discusses the popular interfaces and lists some of the others.

The I/O section of a computer is *not* usually referred to as a(n)

> a. interface
> b. data bus
> c. I/O port
> d. controller

---

19   (*b.* data bus)   Virtually all data transfers between a computer and a peripheral are in the form of 8-bit binary words or bytes. The bytes of data may represent ASCII characters, computer op codes, or just longer binary words that have been divided up into bytes for convenient transmission. The interface transmits the data in byte-sized chunks. The main issues of interest are the format of the data, the speed of the data, and the logic voltage levels used on the CPU bus and in the peripheral circuits.

First, let's talk about the formatting of the data. There are two primary means of formatting data for transmission from one place to another, *parallel* and *serial*. In *parallel data transmission,* all bits of the binary word are transmitted simultaneously. This means that there must be one signal path for each bit of data to be transmitted. See Fig-

(continued next page)

ure 6-4. This is perhaps the most common method of data transfer *inside* a microcomputer. It is convenient since most binary data can be readily transmitted directly over the microcomputer data bus. The primary virtue of a parallel interface is high speed since all bits are moved at the same time. Its disadvantage is that one wire (plus a common ground) must be used for each bit transferred. This limits the distance over which data can be transferred and increases the cost. There are several very popular standard parallel interfaces used in personal computers and embedded controllers which we will discuss later.

Serial data transmission is the other method of data transfer, whereby the bits in a word are transmitted one bit at a time. See Figure 6-5. Such a method is slower than parallel, but many peripheral devices do not often require the very high speed. The primary advantage of the serial method is its simplicity—it requires only one data path, either a wire with a ground or a radio (wireless) link. Many different serial data interfaces are in common use. A modem is a type of serial interface that is used to transmit data over the telephone lines. The modem converts the serial bit voltage levels into audio tones that will pass through the telephone system like voice—more about that later.

Which is faster?

a. serial transmission
b. parallel transmission

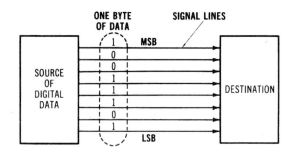

Fig. 6-4. Parallel data transmission.

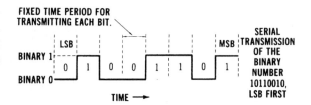

Fig. 6-5. Serial data transfer.

20 (*b.* parallel transmission) Speed is another factor that must be dealt with in an interface. The microcomputer operates at a fixed speed because of its clock. The clock rate can vary, from several megahertz (MHz) in slow microcontrollers up to 200–500 MHz in modern personal computers and network servers. The external peripheral equipment usually operates slower because of its mechanical nature. Further, such devices are not synchronized with the computer, so they operate asynchronously. Typical examples are printers, scanners, bar code readers, keyboards, and robots. Often some kind of speed buffering is required in order to make the two units compatible. This usually means that the faster unit waits for the slower unit, with the slower peripheral sending a signal to the CPU that indicates it is ready for data, has data to send, or is busy and cannot respond. In some cases, the interface may actually contain a small auxiliary memory that will temporarily store data from a fast device and release it at a controlled rate for the slower device.

Another common incompatibility between a CPU bus and a peripheral is logic voltage levels, that is, those voltages used to represent the binary 0's and 1's. The voltage of the logic signals available on the microcomputer data bus typically is not compatible with the signals required by the peripheral unit. The bus is normally set for three-state logic and the peripheral is not. The interface takes care of any logic-level conversions.

Finally, the interface also takes care of the addressing or identification of the external device to be used in a data transfer. Each peripheral is assigned an address, either a unique I/O address or one of the unused memory location addresses as in memory-mapped I/O. The interface contains a decoder that will recognize the address when it is received over the address bus.

Which of the following is *not* a kind of buffering provided by the interface?

    a. addressing
    b. data formatting
    c. logic levels
    d. speed
    e. code conversion

---

21   (*e.* code conversion)   Go to Frame 22.

Common Parallel Interfaces

---

22   Two widely used parallel interfaces are found in personal computers and microcontrollers. These are:

1. Centronics printer port
2. Small Systems Computer Interface (SCSI)

Let's take a closer look at each.

The most common parallel interface is one that is used to connect a printer to a personal computer. It is simply referred to as a *parallel printer port,* or a *Centronics interface* after the printer manufacturer who originally developed it. The parallel interface transfers 8-bits or one byte at a time and includes several status and control lines. It can operate up to a speed of about 50,000 bytes per second (abbreviated 50 kBps—the capital B means byte) over a cable up to 25 feet long. Typical cable lengths are usually less than 10 feet. Over shorter cables, the Centronics interface can be used to transfer data at a rate up to 100 kBps.

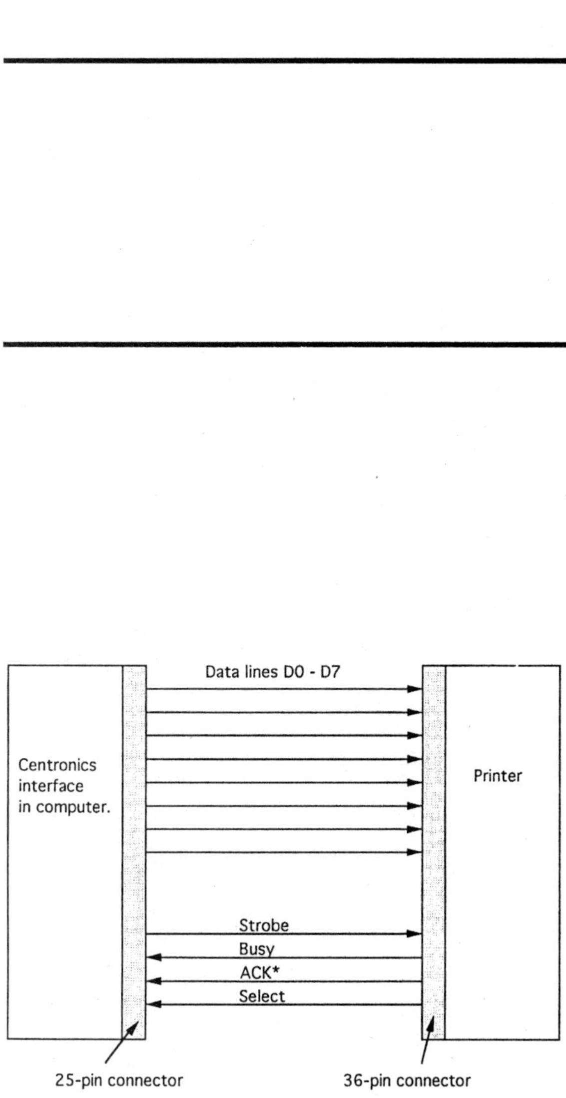

The connections between the parallel interface and the printers are shown in Figure 6-6. The data bits are carried by the D0 to D7 lines, where D0 is the LSB. The data words are usually ASCII characters. The strobe is a pulse that occurs after the data bits are placed on the cable to the printer. The strobe pulse tells the printer to accept the data and store it in a register.

Fig. 6-6. The connections between a PC and a printer using the Centronics or parallel printer interface.

(continued next page)

The ACK* signal is an acknowledgment that the printer sends back to the computer saying that it received the data. Some interfaces do not use it. The BUSY signal generated by the printer tells the computer when the printer is ready to accept data. If it is not, the BUSY line will be high. The SELECT signal is a logic level from the printer that tells the computer that the printer is powered up and on-line ready for service. All logic levels are 0 volts for binary 0 and +5 volts for binary 1.

The cable between the computer and the printer is terminated in connectors. The connector at the computer end is known as a DB25. It has 25 pins, some of which are not used. The other end of the cable uses a Centronics 36-pin connector. Again, some of the pins are not used.

The Centronics interface in most implementations is a unidirectional port; that is, it sends data only from the computer to the peripheral. Further, its primary use is as a printer interface. However, this interface can be made bidirectional so that it can be an input port as well as an output port.

Which signal tells the computer that it is OK to send data?

    a. BUSY = 1
    b. DATA STROBE = 1
    c. BUSY = 0
    d. ACK* = 0

---

23   (*c.* BUSY = 0)   Another popular parallel interface is the Small Systems Computer Interface or SCSI. You will hear it referred to as the "scuzzy" interface. This interface transfers 8-bit bytes between the computer and various peripherals. Its most common use is to connect the computer to magnetic tape backup units, CD-ROM drives, large hard disk drives, or RAID units. A RAID is a "*r*edundant *a*rray of *i*nexpensive *d*isks" or a cluster of hard disk drives set up to store a massive amount of data (many gigabytes or terabytes).

Figure 6-7 shows the basic arrangement of the SCSI. The interface circuitry is plugged into the personal computer and the peripherals are connected to it via a parallel ribbon cable. The cable forms a high-speed bidirectional bus with multiple control lines. The peripheral units are *daisy-chained,* meaning that the bus passes through each peripheral to the other. Data transfers can occur in either direction between two units. The SCSI bus can handle eight units, one being the host computer that is connected to as many as seven peripheral devices, such as disk drives or back-up storage tape drives.

The data rate is up to five megabytes per second (5 Mbps) on the older SCSI-1 interfaces. The newer SCSI-2 standard supports up to 10 MBps of data transfers. By using an additional cable and connector, 16- and 32-bit data transfers can be achieved with a data rate up to 40 MBps.

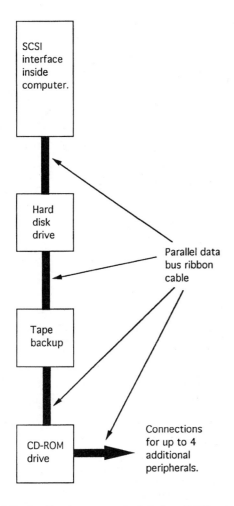

SCSI
interface
inside
computer.

Hard
disk
drive

Parallel data
bus ribbon
cable

Tape
backup

CD-ROM
drive

Connections
for up to 4
additional
peripherals.

Fig. 6-7. Small systems computer interface (SCSI), a typical configuration showing peripherals and ribbon cable interconnects.

With a 32-bit bus, four bytes can be transferred simultaneously at 10 Mbps, giving a maximum of 40 MBps. A SCSI-3 standard defines more than seven peripherals per interface and a single cable and connector. In all SCSI implementations the cable length is limited to six feet to achieve the specified speeds.

SCSI is fast and versatile, but complex. It is a standard interface on some personal computers (Apple Macintosh) but an optional add-on for others.

Which of the following is *not* true about the SCSI?

    a. Its data rate is 5 or 10 MBps.
    b. It is called the scuzzy interface.
    c. It can accommodate up to 15 peripherals.
    d. Its main use is to connect disk drives, tape drives, and CD-ROMs.

24   (*c.* It can accommodate up to 15 peripherals.)
Now let's discuss serial interfaces. Go to Frame 25.

Serial Interfaces

25   Another method of transmitting digital data from one place to another is *serial data transmission*. In serial data transmission, only one signal path is needed since the binary data is transmitted one bit at a time. Refer back to Figure 6-5.

The primary advantage of using the serial method over the parallel method is that only a single path needs to be used between the microprocessor and the external device. This is particularly beneficial for applications for which

(continued next page)

very long connections are required. If parallel interconnections are used, the cost of the interconnecting cable would be roughly eight times that of a serial system, assuming that 8-bit words are used. In addition, there is often difficulty with electrical noise and crosstalk between parallel wires with lengths more than 15 to 20 feet long. *Crosstalk* is the coupling of a signal on one line to another adjacent line in a parallel cable due to magnetic induction or capacitance.

The disadvantage of serial data transmission is low speed. Since each bit is transmitted individually one after another, it takes longer to transmit a word by serial means than by parallel means. A parallel data transfer will actually take place in nanoseconds or, at most, microseconds. On the other hand, serial data transfers require significantly longer time. In most practical computer systems, serial data transfers take milliseconds or, at best, tens of microseconds. This may or may not be a disadvantage, depending upon the application. For example, most computer peripherals are rather slow mechanical devices and, therefore, cannot handle very high speed data transfers. Serial data transfers are fast enough to accommodate such mechanical devices as printers and keyboards. For this reason, the serial interface is used with most external peripheral equipment. Parallel interfaces are used primarily when high-speed data transfers are a must.

The big advantage of serial data transmission is that it can take place over coax cable or two twisted wires, called a twisted pair, for distances of hundreds or even thousands of feet. For this reason, serial transmission is used both for peripheral devices and for long distance data communications such as local area networks (LANs) or remote connection between computers over the telephone system.

The main advantage of serial transfers is:

    a. higher speed
    b. longer distances
    c. lower speed
    d. shorter distances

---

26   (*b.* longer distances)   Serial data transmission is the most widely used method of communication between computers and peripherals. Data is transmitted serially as 8-bit bytes, although other formats are used also. The ASCII code is used almost universally. Remember that the ASCII code includes letters, numbers, and many special command or control characters.

A special format is used for transmitting ASCII data as shown in Figure 6-8. The data is transmitted as a series of marks and spaces that represent binary 1s and 0s respectively. The marks and spaces may be voltage or current pulses, depending upon the circuitry used.

Note in Figure 6-8 that the beginning of a data word transmission is designated by a start bit. Normally, when no data is being transmitted, the serial data line rests at the

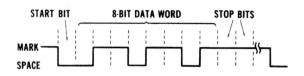

Fig. 6-8. Serial word transmission format for ASCII data.

mark voltage or current level. To signal the beginning of a data word, a *start bit* is sent first. The start bit is a space or binary 0 that lasts for a 1-bit time interval. The 8-bit data word is then transmitted one bit at a time. The data word may be the 7-bit ASCII character, an 8-bit version of ASCII, or the 7-bit ASCII code plus a parity bit for error detection. The end of the transmission is designated by one or, in some systems, two stop bits. One stop bit is the most common.

Refer again to Figure 6-8. Each bit has a fixed time interval. This bit time is referred to as a *baud,* which is the shortest element in the code, in this case one bit. The speed of data transmission is, of course, determined by the time period of the baud. The shorter this time period, the higher the transmission speed. Data speed is expressed as the *baud rate,* which is the number of bits or bauds per second.

Another way to express the term baud is that it represents the number of times per second that the serial line condition (space to mark, mark to space) could change. For most applications, the terms baud and bits per second (bps) are the same and can be used interchangeably. (Note that the lower case *b* means bit, not byte.)

The data rate in bits per second (bps) can be computed with the simple expression:

$$bps = 1/t_b$$

where t_b is the time for one bit. For example, a bit time of 100µS or 100×10^{-6} second is:

$$bps = 1/(100 \times 10^{-6})$$
$$bps = 10,000 \text{ bps or 10k bps}$$
$$(k = 1,000)$$

What is the data rate of a serial bit stream with a bit interval of 104.167 µS?

  a. 2,400 bps
  b. 9,600 bps
  c. 14,400 bps
  d. 28,800 bps

Note: The symbol µS means microsecond, or one millionth of a second.

---

27   (*b.* 9,600 bps)   You can reverse the formula to calculate the bit time for a given data rate:

$$t_b = 1/bps$$

(continued next page)

A popular LAN uses a 10 Mbps data rate. Its bit interval is:

$$t_b = 1/(10 \times 10^6) = .1 \text{ }\mu S \text{ or } 100 \text{ nS}$$

Note: The expression nS means nanosecond, or one billionth of a second.

What is the bit interval of a 56k bps signal?

    a. 14.4 μS
    b. 17.8 μS
    c. 34.77 μS
    d. 69.44 μS

28

(*b.* 17.8 μS)   Data rates vary widely in computer and data communications systems. Most video terminals, printers, and other peripheral devices operate at 9,600 bps. Computer modems operate at 14.4k, 28.8k, 33.6k, or 56k bps. LANs operate at 2.5, 4, 10, 16, 20, and 100 megabits per second (Mbps). Speeds up to 1 Gbps are being used in some networks (*G* means giga, or one billion).

As in parallel data transmission systems, an interface, controller, or adapter, is used to connect the computer to the peripherals for serial data transfers. When serial data peripherals are used, the interface makes the parallel computer compatible with the external serial device. Therefore, one of the main functions of the interface is parallel-to-serial and serial-to-parallel data conversion. The interface also handles all of the other usual functions, such as speed buffering and logic voltage (or current) level conversions.

This serial-to-parallel and parallel-to-serial data conversion is carried out by a special device called a *universal asynchronous receiver transmitter* or *UART*. Another common term is *asynchronous communications interface adapter (ACIA)*. A UART is a single LSI circuit. In addition to performing the serial-to-parallel and parallel-to-serial data conversions, the UART also controls the bit rate, generates the start and stop bits, and provides various control functions.

Figure 6-9 shows a simplified block diagram of a UART. Data to be outputted from the computer to a serial peripheral device is placed on the microcomputer data bus. The read/write control logic in the UART passes the data over the internal UART bus to the transmitter. The word to be sent is stored in a buffer register, and then transferred to a shift register. Start and stop bits are added, and then the data is shifted out one bit at a time at a bit rate determined by the control logic. The shift register actually produces the parallel-to-serial conversion.

Serial data to be input is fed into the UART shift register one bit at a time at the selected bit rate. The data is usually in 8-bit chunks and could be binary or an ASCII code word. After being received, the data byte is stripped of its

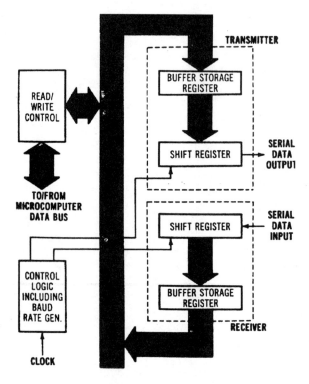

Fig. 6-9. Internal structure of a UART.

start and stop bits and stored in a buffer register. The output of the buffer register is then sent over the UART bus to the read/write control logic that places the data byte on the CPU data bus.

The UART circuit that actually performs the serial-to-parallel and parallel-to-serial data conversion is the

a. shift register
b. storage register
c. bit counter
d. control logic

---

29 (a. shift register)   Although we have illustrated the UART as part of the microcomputer interface, keep in mind that a UART also usually appears inside the serial peripheral device (terminal, printer, etc.) as well. The CPU and the peripheral "talk" to one another via the two UARTs. Also, while many serial interfaces use UARTs, others use special circuitry designed for the specific standard. But in all cases the serial-to-parallel and parallel-to-serial conversion is used.

Serial data transmission is so common that standards have been established to ensure compatibility of interconnection between virtually all computers and peripheral units. The most common serial standard is referred to as the EIA RS232. This Electronic Industries Association standard defines signal voltage levels, control signals, and physical connector sizes. RS232 signal levels are nominally +3 to +25 volts (mark) and −3 to −25 volts (space). The connector is a 25-pin unit containing the single serial data path as well as control lines used for two-way data communications. The connector is commonly referred to as a D connector or a DB25 connector. A 9-pin connector called a DB9 is also used.

The RS-232 interface was originally developed for teletype transmission, but has undergone many changes and enhancements over the years, each marked by a letter suffix on the RS-232 designation. The most recent version is the RS-232E. The RS232 interface can send data reliably at speeds up to 20 kbps for as many as 50 feet.

Figure 6-10 shows the logic level range of an RS-232 interface. Any voltage between 3 and 25 volts can be used. Typical logic levels are +12 volts = binary 0, −12 volts = binary 1. Logic levels of ±5 volts are also common. Signals between ±3 volts are ignored. This helps reduce noise and errors during transmission. This diagram shows the ASCII code for J (01001010) being transmitted. Notice that the LSB is transmitted first (earlier in time).

Which is *not* true about the RS-232?

a. data rate to 20 Mbps
b. distance up to 50 feet
c. logic levels ±6 volts
d. 9 or 25-pin connector

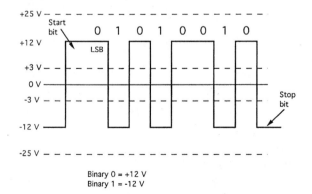

Binary 0 = +12 V
Binary 1 = −12 V

Fig. 6-10. RS-232 signal voltage levels and the transmission of the ASCII character 01001010 representing the letter "J."

30 (*a.* data rate to 20 Mbps) There are many other types of serial interfaces commonly used in personal computers and embedded microcontrollers. Some of the most popular ones are summarized in Table 6-1.

Table 6-1. Common Serial Data Interfaces

Interface	Cabling	Data speed	Applications
EIA RS-449, RS-422, & RS-423	37-pin conn., twisted pair	100 kbps to 10 Mbps	Designed to replace older RS-232 interfaces for greater distances and speeds
Universal serial bus (USB)	4-wire, +5v, & ground, twisted pair for data	12 Mbps max, 1.5 Mbps typical	PC peripherals such as keyboard, mouse, joy stick, printer, audio, and telehone
IEEE 1394 (Firewire)	6-wire, +5v, & ground, twisted pair for data, twisted pair for clocking	100 Mbps min. 400 Mbps max. 800 Mbps (future)	Video applications and multimedia in personal computers
EIA RS-485	Twisted pair cable, differential signals	Up to 10 Mbps	Used in simple networks such as industrial monitoring and control
Ethernet	Coax cable, twisted pair	10 Mbps, 100 Mbps, 1 Gbps	Local area networks
Token Ring (IBM)	Twisted pair	4, 16, 20 Mbps	Local area networks
Fiber Channel	Dual fiber optic cable	133 Mbps to Gbps	Serial interface in very high-speed peripheral devices
Fiber Digital Data Interface (FDDI)	Dual fiber optic cable	100 Mbps, 1Gbps	High-speed backbone for large LANs

As you can see, there are many different types of serial interfaces. The two major categories of interfaces are those for connecting peripheral equipment to a computer and those used for establishing two-way communications in local area networks (LANs). Here is a summary of those listed in Table 6-1.

The RS-449, RS-422, and RS-423 are EIA standards that were established to upgrade and replace the RS-232. The RS-449 standard defines the interconnections via a 37-pin connector. The RS-422 and RS-423 standards define the electrical specifications. The RS-423 is a single-ended interface like the RS-232 in which one of the transmission lines is ground. It restricts the logic levels to six volts maximum. Data rates can be up to 100 Kbps up to a distance of 4,000 feet (a considerable improvement over the RS-232). The RS-422 version specifies a balanced or differential line with logic levels of less than two volts. With this arrangement, data speed can be as much as 10 Mbps at up to 4,000 feet. These interfaces also make it possible to connect to more than one peripheral device at a time, unlike the single connection of the RS-232.

But despite the improvements over the RS-232, these interfaces have not become popular. Instead, the RS-232 continues to be popular. However, although it is a very widely used serial interface it is quickly being replaced by the universal serial bus (USB) in new personal computers for keyboards, printers, the mouse, and other devices such as bar code readers and scanners. It is expected that the USB will be more widely adopted in the future.

The IEEE 1394 is a very high-speed serial interface that was designed to handle the transfer of high-speed digital video data. You will see it in future computers and in consumer electronic devices as more video is implemented in digital rather than analog form. It will be used in digital video cameras, VCRs, and in multimedia applications with digital sound and video on CD-ROMs or the newer high-density digital versatile disks (DVD).

The other interfaces listed in Table 6-1 are used in LANs. By far the most popular is Ethernet with twisted pair. Probably over 80 percent of all LANs use some variation of Ethernet. Next in popularity to Ethernet is Token Ring, IBM's LAN interface. The EIA RS-485 uses a variation of the RS-422 interface in a bus arrangement for bidirectional communications. It is used in industrial networks for monitoring and control operations.

The USB is used primarily in

  a. peripheral interconnections
  b. local area networks

---

31 (*a.* peripheral interconnections) Now answer the Self-Test Review Questions on the accompanying disk.

Embedded Microcontrollers

LEARNING OBJECTIVES

When you complete this unit, you will be able to:

1. Explain the difference between the Von Neumann and Harvard architectures as used in modern microcontrollers.

2. Name four of the most popular embedded commercial microcontroller chips.

3. Describe the specifications and special features of the Intel MCS51 microcontroller series.

4. Describe the specifications and special features of the Motorola 68HC11 microcontroller.

5. Describe the specifications and special features of the Microchip Technology 16C5X series of microcontrollers.

6. Explain the concepts and operation of a digital signal processor and name some common applications.

7. Describe the specifications and special features of the Texas Instruments TMS320C25 digital signal processor.

Microcontroller Architecture

1 The hardware organization of the processors we have been discussing so far is generally known as the Von Neumann architecture. John Von Neumann, a Hungarian-born, Princeton University mathematician, is generally credited with putting together the concept for the stored-program digital computer during and just after World War II. The basic features of Von Neumann architecture are (1) both instructions and data are stored in a single memory address space, and a program counter is used to keep track

(continued next page)

of the current and next instruction to be executed and (2) in the fetch-execute process, both instruction and data words pass over a single CPU data bus.

Many computers still use the Von Neumann organization, but it does become a disadvantage when very high speed operations are needed. All computer operations are serialized in that instructions are fetched and then executed in sequence and the execution often requires the CPU to obtain data from memory prior to fetching the next instruction in sequence. As programs become longer and more complex, operations slow down because of this sequential access of instructions and data from a single memory space. For very large and complex problems, computer programs may no longer seem to run instantaneously. In fact, they may actually take not only many seconds, but in some cases minutes or hours to run completely.

The obvious way to overcome this problem is to use a faster CPU running at a higher clock speed and faster memory chips. Advancements in semiconductor switching speeds over the years have made it possible to continue to use the Von Neuman architecture in many computers. But at some point, because of the application or the computer hardware itself, the Von Neumann arrangement becomes a major limitation. It is usually referred to as the "Von Neumann bottleneck."

One solution to this problem is to incorporate what is often referred to as Harvard architecture. A computer using Harvard architecture has two memory spaces, one for the program and another for the data. The CPU is designed to access instructions from a separate program memory and to get the data from a separate data memory. Each has its own access bus. In this way, both instructions and data can be accessed at the same time if necessary. Such parallel accesses greatly speed up operations.

Figure 7-1 shows the arrangement of a Harvard architecture computer. Note the dual independent paths for data and instructions. Most modern single-chip microcontrollers use some form of Harvard architecture. The instructions are usually stored in an on-chip PROM or EPROM while data is stored in a small on-chip RAM or register array. Some of the single-chip microcontrollers to be discussed in this unit use this architecture.

A Von Neumann architecture computer has:

    a. independent program and data memories
    b. two CPUs
    c. a single memory for both instructions and data
    d. a high-speed advantage over Harvard architecture computers

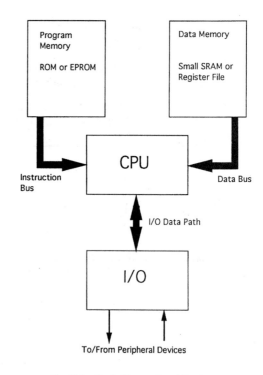

Fig. 7-1. Basic Harvard architecture.

---

2   (*c.* a single memory for both instructions and data)
Go to Frame 3.

3   Intel was the inventor and manufacturer of the first practical commercial microprocessor in 1971 and today is still the leader in microprocessors. It is the world's largest semiconductor manufacturer. Although Intel is most well known for its popular 8086/286/386/486 and Pentium microprocessors, as used in over 85 percent of the world's personal computers, Intel is also a manufacturer of embedded microcontrollers. Their first single-chip microcomputers were known as the MCS-48 series. This 8-bit microcontroller was one of the first to incorporate CPU, memory, and I/O circuits on a single chip. It was widely used in many products through the late 1970s and 1980s.

A more recent improved and updated version is the MCS-51 series. Also an 8-bit microcontroller for embedded applications, it offers improved features and performance. Over the years it has been built into many consumer, commercial, and industrial products. Still popular today, many newer and faster versions, some with a 16-bit word format, are available. Variations of this chip are also available from second sources such as Phillips Semiconductor and Dallas Semiconductor. This section summarizes the features and specifications of this microcontroller.

Figure 7-2 is a simplified block diagram of the MCS-51 series devices. The basic model has the part number 8051, and this series is referred to by this designation. The original 8051 was made with Intel's high-speed N-type MOS-

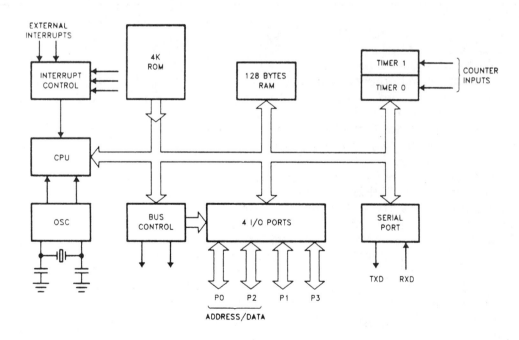

Fig. 7-2. Basic architecture of the 8051. *Courtesy Intel Corp.*

(continued next page)

FET circuits called HMOS; however, more recent versions use smaller and faster CMOS circuits for improved performance. The 8051 features an 8-bit CPU, Harvard architecture with an on-chip program memory (ROM) of 4k bytes, and a data memory (RAM) of 128 bytes. Different versions of the MCS-51 have other RAM and ROM sizes. There are versions with 4K, 8K, or 16K of masked ROM. There are also versions with similar amounts of EPROM. The EPROM versions have ultraviolet erase capability, making it possible to reprogram the chip. These versions are good for development prototypes for which it is often necessary to add code or fix bugs. The EPROM versions can also be used in products for which it is desirable or necessary to upgrade at a later date. A version of the 8051 is also available with no on-chip ROM of any type. An external ROM is used to contain the program. To address off-chip memory, I/O ports P0 and P2 are configured for output and act as a 16-bit memory address bus. Up to 65,536 bytes (64KB) can be addressed.

Although the basic version of the 8051 has 128 bytes of on-chip RAM, others have 256 bytes. If more RAM is needed, it must be added off-chip. I/O ports P0 and P2 are used as a 16-bit address bus to address up to 64KB of external RAM if needed.

Which of the following is true? All 8051 versions have:

a. EPROM
b. off-chip RAM and ROM
c. on-chip ROM and off-chip RAM
d. on-chip RAM

---

4 (d. on-chip RAM)   A more detailed block diagram of the 8051 is given in Figure 7-3. The RAM and EPROM/ROM blocks are shown. Note that there is a RAM address register and a separate program address register. The program address register gets its input from the program counter, a program counter incrementer, a buffer register, or the data pointer register (DPTR). Note the stack pointer that is used to point to a stack implemented in RAM.

The CPU features a single 8-bit accumulator and two temporary registers (TMP1 and TMP2) that hold operands during the various operations. The ALU performs a wide variety of arithmetic and logical operations. The 8051 does have multiply and divide instructions. The B register is used with the accumulator during the multiply and divide operations. The CPU also has an 8-bit program status word (PSW) register that contains flag bits that are set or reset as the result of CPU operations. These include carry, overflow, and general user flags used by the jump instructions for decision-making purposes.

The 8051 also has a set of special function registers (SFRs). These registers are a part of the on-chip RAM. The RAM from 00H to 7FH (remember that H designates a hex

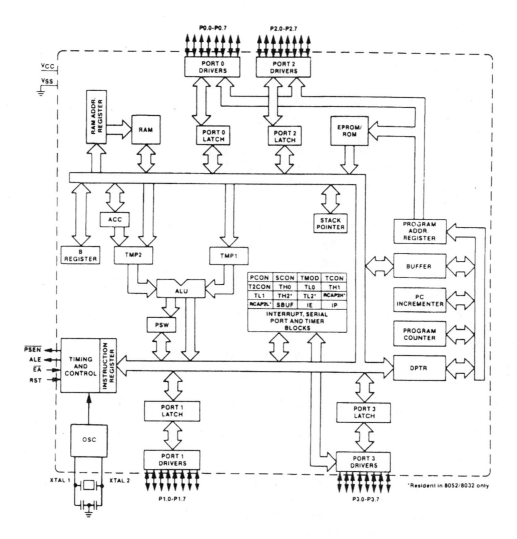

Fig. 7-3. Detailed organization of the 8051. *Courtesy Intel Corp.*

number or address) is the 128-byte RAM area for general computing. RAM addresses 80H to FFH contain the SFRs. SFRs are a mix of operational registers, I/O registers, and control registers. The accumulator, B register, PC, stack pointer, PSW, and DPTR are duplicated and defined by addresses in the SFR area. The SFRs also include the I/O port data in and out registers, timer/counter registers, and a variety of control registers that set up the I/O ports, timers, and power control.

SFRs can be addressed as RAM.

    a. True
    b. False

5   (*a.* True)   As for input-output capabilities, the MCS-51 offers four 8-bit parallel bidirectional I/O ports and a full duplex serial I/O port incorporating its own

(continued next page)

UART for communications. Refer back to Figure 7-3. A separate interrupt control system permits from 5 to 16 separate hardware interrupts from external sources depending upon the specific method.

The I/O ports can be configured as either inputs or outputs. The input and output data registers associated with the I/O ports are contained within the SFR area of RAM. Since these registers are treated as RAM and have addresses, the I/O operations are of the memory-mapped I/O variety. Any of the general-purpose data transfer instructions (move or MOV instructions in the 8051) can be used in I/O operations.

The 8051 also contains a serial I/O port. The port is full duplex, meaning that it can both send and receive serial data simultaneously. The port may be configured for several modes of operation, including both synchronous or asynchronous transmission. In synchronous mode, the 8-bit words are transmitted end to end, LSB first, at a data rate set by the clock frequency and a control word. Data rates up to one Mbps are possible.

In asynchronous mode, both start and stop bits are used at the beginning and end of the word. A programming option permits an extra 9th data bit to be included for parity or other purposes. The LSB is transmitted first, the 9th bit last. The data rate is programmable.

The serial port uses port 3 pins for input and output. Pin 0 is used to receive the serial data while pin 1 is used to transmit serial data. Separate shift registers, not shown in Figure 7-3, are used to receive the serial data and to store it for transmission. However, all serial data must pass through the serial data buffer (SBUF) register in the SFR area. Data to be output are stored in SBUF and then transmitted to the serial output shift register. Any received data is transferred from the receive shift register to the SBUF for use by the program.

Serial I/O operations are set up using the serial port control register (SCON) located in the SFR RAM area. It contains bits that define the mode, baud rate, and interrupts.

A special feature of the MCS-51 is the use of on-chip timers. Although timing and delay operations may be programmed, the separate timers make such operations faster and more convenient. The 8051 has two 16-bit counters designated T0 and T1 that can be used to count external events or an internal clock for timing operations.

In the counter function, the external input to be counter is applied to port 3 pin 4 for T0 and pin 5 for T1. The counter increments on the 1-to-0 signal transition. The maximum count is 65,535 for each counter. When the 65,536th input is received, the counter rolls over to zero and begins again. In the timer function, several modes of operation are possible. In all modes, the counters are set up to count the clock signal divided by 12. If the clock is 12 MHz, the counters count a 1 MHz timer clock signal, thus giving the timer 1 mS resolution. With 16-bit counter registers, the timers can produce timing delays up to

65,536 mS. The timers can be set up as 8- or 16-bit counters with or without a divide-by-32 prescaler. The prescaler is a 5-bit counter/divider connected between the timer clock and the input to the timer counter.

In both counter and timer modes, the register contents of the counters appear in the SFR area of RAM. Timer T0 has two bytes designated TL0 and TH0 for the low and high bits of the 16-bit word. Timer T1 has two bytes designated TL1 and TH1 for the low and high bits of the 16-bit word. These bytes may be preset to specific values so that special count times can be programmed. The timers generate outputs when the counters overflow and these outputs trigger interrupts.

The timers and counters are set up using the Timer/Counter Mode Control Register (TMOD) and the Timer/Counter Control Register (TCON), both in the SFR group. The mode and configuration is specified by two 8-bit words that are stored in TMOD and TCON registers at the beginning of a program.

When the 8051 counters overflow, they:

    a. generate delays
    b. generate an interrupt
    c. give an incorrect count
    d. stop counting

---

6   (*b.* generate an interrupt)   The instruction set for the 8051 contains many flexible data transfer instruction, arithmetic, and logical operations, as well as program branching instructions. The data transfer instructions have the mnemonic MOV for move and can be used for just about any conceivable register-to-register, register-to-memory, or memory-to-register transfer. The addressing modes include direct, indirect, immediate, relative, and index operations discussed earlier. In the direct mode, the data address is the second byte of the instruction. Any word in the RAM or SFR area can be accessed. In the indirect mode, the address is in the stack pointer or in the 16-bit data pointer register (DPTR).

A register address scheme is also used by some instructions. The RAM area from address 00H to 1FH, a total of 32 bytes, is divided up into four register banks designated 0 through 3. Each bank contains eight bytes. Some instructions can designate a specific register in a specific bank with the codes in the second byte of the instruction.

The relative mode of addressing is used with the branch instructions. The second byte of a jump instruction contains an 8-bit 2's complement number representing values from 2^7 to 2^{-8} or +127 to −128. This value is added to the content of the program counter (PC) to form the address to which the CPU will get the next instruction. The branch address is relative to the current instruction address. The

(continued next page)

instruction to be executed next lies within a range +127 bytes beyond the address of the branch instruction or backwards 128 bytes from the branch instruction. An example illustrates this idea. See Figure 7-4.

Assume that the program contains a branch instruction stored at memory address 70 decimal or 46 hex. In the second byte of the instruction is an offset value, in this case −28 decimal or E4 hex. If the condition of the branch instruction is met, the CPU will branch to the address formed by adding the content of the program counter, which contains the address of the branch instruction just addressed (70), and the offset value (−28). The result is 70 − 28 = 42 or A2 hex. The next instruction will be fetched from RAM location 42. Note that with this arrangement that you can branch backwards to form a loop from the branch instruction address ADR to ADR − 128 or branch forward to ADR + 127.

If a branch instruction is located at address 55 decimal and the offset is +18, and the branch condition is met, the next instruction will be taken from address

    a. 37
    b. 57
    c. 73

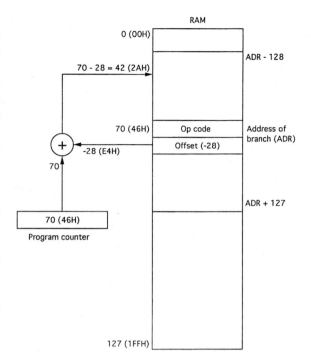

Fig. 7-4. An example of computing a relative address.

7

(*c.* 73)   One final feature of the MCS-51 series is that in some of the CMOS versions, a power-saving scheme is included to minimize the current drawn by the device when it is not being used. This allows the 8051 to be used in some battery-powered applications. To lengthen battery life, the microcontroller can be programmed to automatically turn itself off, thus greatly reducing the current drain on the battery.

The 8051 has two power-saving modes, idle and power-down. Either mode is entered by storing the appropriate binary code in the power control register (PCON) that is part of the SFRs. In the idle mode, power is cut off to most of the circuitry except the clock, timer, and serial port. The CPU is not clocked, but all register contents are preserved. The 8051 comes out of the idle mode if an interrupt occurs. The idle mode can also be terminated by an external hardware reset on the 8051 reset pin.

The other power-saving mode is power down. In this mode, even lower current drain is obtained. The clock is stopped completely, but the SFRs and all RAM contents are preserved. The supply voltage may be reduced to as low as two volts to further reduce battery drain. A hardware reset is needed to bring the 8051 back to life.

The power-savings feature of the 8051 is that it

    a. permits smaller power supplies
    b. is used when the CPU is not that busy
    c. allows batteries to be used as the main power supply
    d. is available on all versions of the MCS-51

8 (*c.* allows batteries to be used as the main power supply) Go to Frame 9.

Motorola 68HC11 Microcontroller

9 Motorola introduced its first 8-bit microprocessor (6800) back in the mid-1970's. It immediately became popular and since then has been widely adopted for many embedded applications. Further, over the years many different variations and offshoots have been created following the original basic design. Today Motorola markets the 68HCXX series of microprocessors, which are among the most widely used in the world. These single-chip microcomputers contain a complete CPU, RAM, PROM, or EPROM, plus a variety of I/O circuits. These microcontrollers are widely used in automotive applications and in many consumer electronic products. One of the most popular variations is the 68HC11 that we will discuss here.

Figure 7-5 is a general block diagram showing all the various features of the 68HC11. It has an 8-bit CPU, 128 or 256 bytes of on-chip RAM, PROM or EPROM whose actual size depends upon the version of the device, and five I/O ports. The I/O circuitry also includes programmable timers and an A to D converter.

The basic register architecture of the 68HC11 CPU is shown in Figure 7-6. It contains two 8-bit accumulators designated A and B. These can also be used as a single 16-bit accumulator designated as D. Also included are two index registers designated X and Y. The basic architecture also includes a 16-bit program counter, a 16-bit stack register, and an 8-bit condition code register (CCR) that monitors the status of the CPU.

The basic architecture of the 68HC11 is of the Von Neuman type, which uses a single address space for both RAM and ROM. Memory mapped I/O is also a standard feature. With a 16-bit program counter, the total addressable memory space is 65,536 bytes or 64KB. The first 256 bytes of memory ($00 to $FF, where the prefix $ is used by Motorola to designate a hex address or data word) are reserved for RAM that is on the chip. ROM or EPROM is usually located in the region of memory from $D000 to $FFFF. The memory space from $1000 to $103F is allocated to a 64-byte register block. These on-chip registers are assigned to the I/O ports, timers, and counters as well as a variety of control functions.

The CPU has a large instruction repertoire that includes a variety of load-and-store instructions that permit data to

(continued next page)

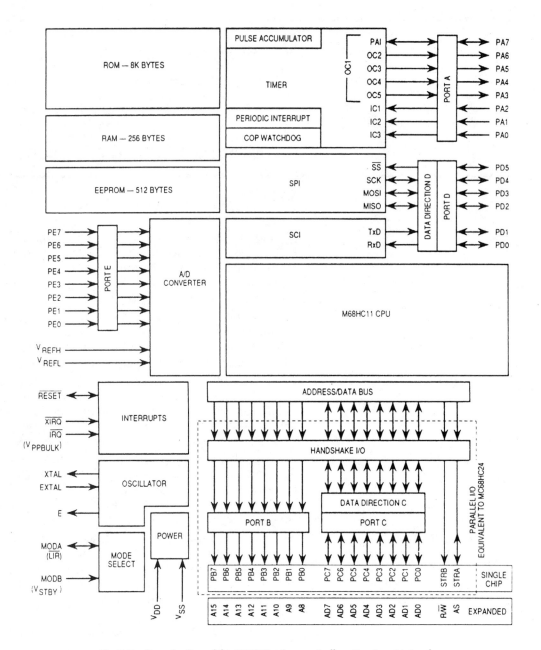

Fig. 7-5. Organization of the 68HC11 microcontroller. *Courtesy Motorola.*

be moved between registers and memory. The arithmetic instructions include both multiply and divide instructions, which eliminate the need for these operations to be programmed as is necessary with many embedded controllers. Boolean logic instructions are also included.

The 68HC11 contains a variety of branch instructions that permit decisions to be made based upon a variety of conditions, including zero, minus, overflow, and a number of other variations. Branching instructions use a relative addressing scheme that permits them to jump forward or backward in the program over a +127 to –128 byte memory range. The relative addressing mode is reserved for the

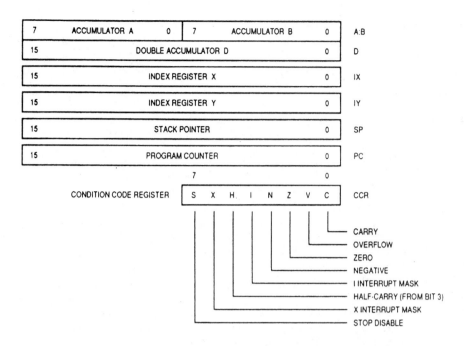

Fig. 7-6. Programming registers in the 68HC11. *Courtesy Motorola.*

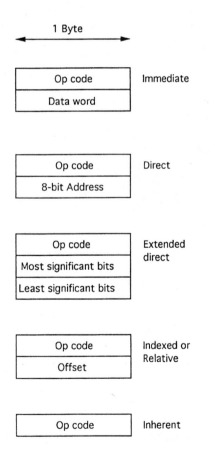

Fig. 7-7. Instruction word formats of the 68HC11.

branching instructions and is similar to that used in the 8051.

The basic instruction word formats are shown in Figure 7-7. The addressing modes include immediate, direct addressing with an 8-bit address; extended direct addressing with a 16-bit address; and indexed addressing, relative addressing, and inherent addressing. The first byte of an instruction contains the op code defining the operation. For immediate instructions, the second byte is the 8-bit data word. The immediate instruction may also be used with a 16-bit data word by placing the most significant eight bits in the second byte of the instruction and the least significant eight bits in the third byte of the instruction.

For direct addressing, the second byte of the instruction is an 8-bit address that defines a location in the first 256 bytes of on-chip RAM. Extended addressing uses the second and third bytes to contain a 16-bit address with the most significant eight bits in the second byte and the least significant eight bits in the third byte. For indexed addressing, the second byte of the instruction contains an 8-bit offset value that is added to the content of the designated index register to form the address of the operand. For branching instructions, the second byte is an 8-bit two's complement signed number defining a value to be added to the content of the program counter to establish the branch address. Inherent addressing uses a single byte instruction. The op code implies which data word and its location (usually a register) to use in the operation.

(continued next page)

The 68HC11 uses Harvard architecture.

a. True
b. False

---

10 (*b.* False)   The 68HC11 has an incredibly versatile I/O system. It consists of three 8-bit parallel ports designated A, B, and C; a fourth port dedicated to serial I/O operations; and a fifth port designated E, which is used for analog signal multiplexing and A-to-D conversion. The I/O functions also include a versatile programmable counter/timer.

As indicated earlier, the I/O system uses a 64-byte block of the memory space starting at hex location $1000. These 64 locations designate input/output registers and a variety of special set-up and control registers are used to establish the function of each I/O operation.

I/O port A has three dedicated input lines and three dedicated output lines, and two lines that can be programmed as either inputs or outputs. These lines can be used where only a few bits of I/O data or I/O signals are needed. Some of these lines may also be configured as input lines to the counter timer.

I/O port B is an 8-bit output-only port. It has the address $1004. Port C is an 8-bit input or output port with the address $1003. The I/O lines are fully programmable in that all lines may be used for input, all for output, or any combination of input or output lines. To set up Port C, load an appropriate bit pattern into the data direction register (DDRC), which is the address $1007. To make a pin an input, store a binary 0 in the correct location. Storing a binary 1 in the correct location designates an output pin. For example, to set up the four least-significant bits as outputs and the four most-significant bits as inputs, store the value $0F in the DDRC.

Port D contains I/O lines associated with the serial interface functions built into the 68HC11. These two systems are the serial communications interface (SCI) and the serial peripheral interface (SPI). The SCI is a standard asynchronous serial communications port that permits full duplex input/output operations using the standard asynchronous data format. Data is transmitted in 8-bit groups along with a start and stop bit. The SCI may be programmed to transmit a 9-bit word, eight for data and an additional for parity.

The SPI interface is a synchronous data port that transmits 8-bit data words serially end to end. In both the SCI and SPI, the serial baud rate is fully programmable from approximately 150 bps to 125 Kbps.

The 68HC11 contains an extensive timer/counter section for generating and measuring time intervals, generating delays, or for counting external or internal events. The timers are driven by an internal clock signal designated the E-clock, which is derived by dividing the CPU clock by 4. For an 8 MHz system clock, the E-clock is 2 MHz. The E-clock also drives a prescaler, which is a selectable frequency divider that can divide by 1, 4, 8, or 16. The prescaler output drives a 16-bit counter designated TCNT in the register area of RAM.

The timer/counter has three input capture counters for event counting and five counters that can be compared to register values set up during programming. The specific timer operation is specified and set up during programming by setting the correct bit patterns into the timer mask and control registers. As in other microcontroller timers, operations usually result in the generation of interrupts to tell when a specific time duration or count value has been achieved.

A unique feature of the 68HC11 is its built-in analog-to-digital converter (ADC). An 8-bit successive-approximations ADC allows external analog signals to be digitized and used as inputs to the microcontroller. External reference pins let you set the analog voltage range over which the ADC will recognize and convert. Most applications simply use the 68HC11's power source (+5 volts and ground) as the reference inputs. Analog signals in the 0 to +5 volt range may be converted to some binary value between 00000000 and 11111111 with an 8-bit resolution. With 8-bits the resolution is one in 256 or 1/256 = .0039. The minimum voltage increment is 5/256 = .01953 volt or 19.53 mV. The ADC converts the analog input to the closest digital value within ±19.53 mV.

The ADC also has a built-in 8-input analog multiplexer. This allows you to sample up to eight different analog inputs. The inputs appear at the eight port E inputs. The channel to be digitized is selected by the ADC control word ADCTL stored at RAM location $1030. The ADC output is stored in one of four registers ADR1–ADR4 designated with addresses $1031–$1034.

The ADC feature is truly handy in many embedded applications. Many times it is desirable to monitor some analog variable such as temperature, pressure, physical position, or light level. The physical variable, whatever it may be, is detected, sensed, and measured by a sensor or transducer that converts it into an electrical signal. This analog signal is filtered, amplified, or otherwise conditioned and applied to the multiplexer input for data conversion. In this way, analog data may be used as part of the controller program. It is especially useful in closed-loop control applications in which an analog signal provides the feedback used to make decisions about what outputs to generate.

How does the CPU access the digital value of the analog input?

a. It is in the accumulator.
b. It is stored in a RAM address designated by an instruction.
c. The multiplexer address determines the data location.
d. The value is in one of the ADR registers.

11 (*d.* The value is in one of the ADR registers.)
Go to Frame 12.

Microchip Technology PIC16C5X

12 Perhaps the smallest and least expensive embedded microcontroller chip on the market today comes from a company called Microchip Technology, Inc. They specialize in making ultrasimple and inexpensive single-chip microcontrollers for embedded applications. The PIC16C5X is representative of a whole series of single-chip 8-bit microcontrollers made by this company. In very large quantities, these microchips are less than $2 apiece, making them suitable for even the smallest and cost-conscious applications. They can be used in consumer appliances, remote controls for infrared and RF applications, or automotive control.

The PIC16C5X microcontrollers are 8-bit devices made with CMOS circuits. They feature a Harvard architecture with separate program and data memories that have separate internal busses. This allows simultaneous instruction and data feed operations that significantly speed up all operations. The maximum clock speed is 20 MHz, although most applications use a lower frequency clock. The device is available in 18- and 28-pin DIP configurations. Different versions of the PIC16C5X are available; the difference between models is in the RAM, ROM, and EPROM capacities. A primary feature of the PIC16C5X is its extremely low power consumption, making it ideal for battery-operated applications.

Figure 7-8 shows a general block diagram of the PIC16C56 microcontroller. The major architectural circuits are shown as blocks being interconnected with various data and control lines. The dark lines represent busses of multiple connections. To simplify the drawing of the diagram, a bus is designated as a heavy line and an arrow pointing in the direction of the data transfer. The number of individual lines in the bus is designated by a slash through the line and an adjacent number. A slash with an 8 adjacent to it means that the heavy line represents eight individual lines.

The program memory is either EPROM or ROM. In some versions of the PIC16C5X, the program memory may be masked ROM, UV erasable EPROM, or a special programmable read-only memory that can be programmed only once. This one-time programmable (OTP) ROM makes programming the microcomputer from a personal computer fast and simple. Some versions permit electrical programming of the EPROM from an external computer via a

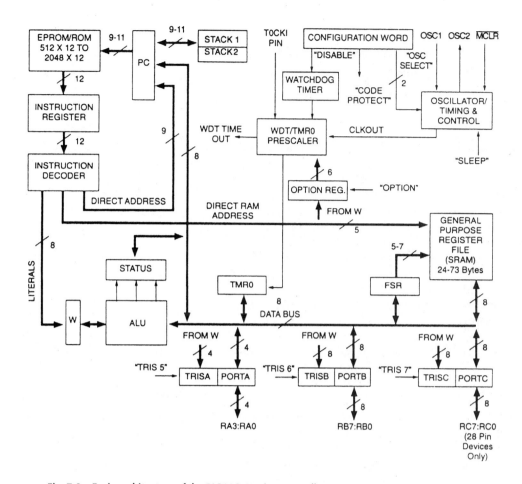

Fig. 7-8. Basic architecture of the PIC16C5X microcontroller. *Courtesy Microchip Technology.*

serial download. The EPROM can be used again by programming over a previously written program.

The program memory can store anywhere from 512 to 2,048 12-bit instruction words. The program memory is organized in pages of 512 words each. The first page, designated page 0, is on-chip. The other three pages of memory may be added externally as required to accommodate the program size.

The PIC16C56 does not use RAM as such. Instead, its data memory is simply a large general-purpose register file. In some versions of the chip, 25 8-bit registers are available for data storage. In other versions, 72 or 73 bytes of data RAM are available. Overall, the data memory consists of up to 80 8-bit registers on-chip. Thirty-two of these registers are considered to be working RAM and can be addressed directly by instructions containing a 5-bit register address. This area is called the *general-purpose register file*. Another area of the data memory is designated as special function registers (SFRs). The SFRs include the I/O port registers, timer/counter registers, program counter, status

(continued next page)

register, and the file select register (FSR). The FSR is used by some instructions to indirectly address the first 32 bytes or RAM (the general-purpose registers).

The PIC16C5X uses which form of architecture?

    a. Harvard
    b. Von Neumann

---

13   (*a.* Harvard)   The instruction set for the PIC16C5X contains a total of only 33 instructions. Despite its simplicity, this instruction set can be used to perform virtually any operation that any larger or more complex microprocessor can perform. Unlike most other 8-bit microcontrollers, the PIC16C5X uses a special 12-bit instruction word. All instructions regardless of the addressing mode use this single 12-bit instruction word format. References to data locations are contained in the 12-bit op code. Other 8-bit microprocessors, as you have seen, have 8-bit instructions that require multiple bytes for some operations. Each of instructions can be executed in a single clock cycle (200nS at 20 MHz clock frequency).

As in any computer, a program in the PIC16C5X is executed one instruction at a time. Each instruction in the program is fetched sequentially from the program memory. The instruction is brought into the instruction register decoded to determine which instruction is to be executed. Signals are generated and sent to other parts of the microcontroller to carry out the operation.

Refer again to Figure 7-8 and note that the PIC16C5X contains a basic ALU. The main computation register is the W, or working, register. Instructions involving two operands or words to be processed access one word from the W register, while the other is in a location designated by an address that is part of the instruction. The second operand is usually contained within one of the general-purpose registers that make up the data RAM.

As instructions are executed by the ALU, bits or flags in the status register are set or reset to indicate a carry or zero condition. These bits are tested by some instructions to provide decision-making operations.

Input/output operations are provided by two or three I/O ports depending upon the specific version of the PIC16C5X. Eighteen-pin versions of the microcontroller have only two I/O ports. I/O port A has four bits while port B has eight bits. Any of these bits may be configured as either inputs or outputs. Twenty-eight-pin versions of the microcontroller have one additional bit I/O port, port C.

The PIC16C5X also contains a timer circuit. It features an 8-bit counter register that can be incremented externally or internally. This counter is both readable and writable under programmable control. The timer also contains an 8-bit software prescaler. The clock speed for the internal counter or prescaler is fully selectable. As with timers and other controllers, this one can be used for a variety of counting, timing, and delay operations necessary when interfacing to external equipment.

The PIC16C5X also has a power down mode. All register and RAM contents are maintained as are I/O states, but the clock is stopped and nonessential circuits are disabled. This feature lets you to disable the controller when it is not in use. This will greatly reduce power consumption and permits the PIC16C5X to be used in virtually any battery-powered application.

To enter the power down mode, a SLEEP instruction is executed. During the "sleep" period, the processor draws only a few microamperes of current. To bring the controller out of the sleep mode, an external reset signal must be applied. The sleep mode can also be abandoned by an on-chip watchdog timer. This on-chip timer circuit runs at a frequency set by values of an internal resistor and capacitor combination. The basic timing period is 18 mS, but with the use of an internal prescaler, this time can be extended to about 2.3 seconds. The watchdog timer enables you to time how long the controller is put to sleep.

The length of a PIC16C5X instruction word is

    a. 8 bits
    b. 12 bits
    c. 16 bits
    d. depends upon the type of instruction

---

14   (*b.* 12 bits)   Go to Frame 15.

Digital Signal Processing (DSP) and the Texas Instruments TMS32O Series DSP

---

15   A smaller yet rapidly growing class of embedded microcontroller is the *digital signal processor*. This special microcomputer is designed to carry out operations of a signal processing nature but in digital form. We normally think of electronic signals to be of the analog variety, which are processed by some type of electronic circuit. Typical processing operations include filtering, phase shifting, modulation and demodulation, and spectrum analysis. In the past, such processes were carried out by analog circuits specifically designed for the purpose. All of these processing operations can be defined in mathematical terms. This being the case, the resulting equations can be

(continued next page)

solved on a digital computer, specifically a single-chip microcomputer. This is called *digital signal processing (DSP)*.

The basic concept of DSP is simple. Take an analog signal, digitize it with an analog-to-digital converter (ADC) into a stream of binary numbers representing amplitude values over time, and then store these numbers in a memory. Use a microcomputer programmed to do the desired math to implement a specific processing procedure. Process the captured signal in digital form and store the result. Finally, output the digital data developed as the result of processing to a digital-to-analog converter (DAC) to end up with an analog output signal. The overall processing effect is similar to what you could achieve with an equivalent analog circuit.

Upon initial consideration, DSP doesn't make sense. What is the value in replacing simple, low-cost analog (linear) processing circuits with complicated data conversion circuits, a costly digital computer, and complex mathematical equations? First, today data conversion (ADC and DAC) circuits are simple and inexpensive. This also holds true for the microcomputer. Special high-speed microcontrollers have been developed especially for DSP, and their cost has declined over the years. Second, DSP produces results that are superior to equivalent analog circuits. For instance, DSP filters produce sharper, more selective circuits with improved performance over comparable analog filters. This is true for many other circuits. Third, DSP produces new processing options that are not available in analog form. A good example is spectrum analysis. In spectrum analysis, a complex input signal—digital or analog—is subjected to a mathematical process called the Fast Fourier Transform (FFT). The process divides a signal into its elementary components which, according to the Fourier theory, are harmonically related sine and/or cosine waves of different amplitudes and phases.

Figure 7-9 shows a complete DSP system. The analog signal to be processed is applied to an analog-to-digital converter (ADC). The ADC samples or measures the signal multiple times as it occurs. For each sample, the ADC generates a binary number proportional to the amplitude of the input signal. As a result, the analog signal is translated into a sequence of binary numbers that is usually stored in a RAM to be processed. If the processing can be accomplished fast enough, then no storage is needed.

The most important consideration of the ADC portion of the DSP is the sampling rate of the signal. In order to preserve the information in an analog signal, the signal must be sampled a minimum of two times for the highest frequency component. For example, if the input is voice from a microphone containing frequencies up to 3 kHz, then the sampling rate must be at least twice that, or 6 kHz. In practice, the sampling rate is usually a bit more than the minimum of two times. The DSP systems used in digital telephones sample the voice signal at an 8 kHz rate. This produces one binary number every 1/8000 = .000125 sec-

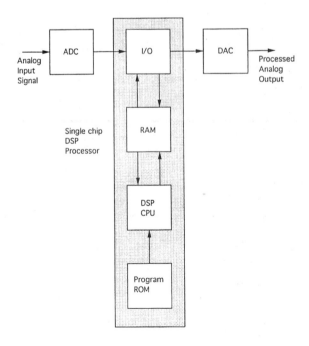

Fig. 7-9. General block diagram of a digital signal processor.

ond or 125 µS. That's 8,000 samples every second. You can see that digitizing data produces a great deal of binary data. One minute of voice at an 8 kHz rate would produce 480,000 binary numbers! A high percentage of ADCs in DSP systems use 8-bits. However, longer word length ADCs can be used (10, 12, or 16 bits) if higher resolution signal representation is required of the application.

The data produced by the ADC represents the voice or other analog input. This data is then subjected to processing by a special microcomputer that has been optimized for DSP. The algorithm implemented by the program, usually stored in a separate program ROM, is one or more higher-level math techniques beyond the scope of this book. The algorithm usually implements one processing scheme such as filtering, although several algorithms may also be used if required. A good example of a multiple algorithm DSP application is a modem. Virtually all modern modems (to be covered later in Unit 12) use a single DSP chip to perform all of the filtering, modulation, demodulation, data compression and decompression, and other processes required.

The processing produces new digital output data. This is stored in a RAM or, if the processor is fast enough, it simply outputs a stream of binary data. This data is fed to a digital-to-analog converter (DAC) that produces the processed analog output signal.

It is important to point out that the circuit in Figure 7-9 can be divided up in several different ways. For example, the ultimate DSP system contains all of the circuitry shown on a single chip. In other applications, the ADC and DAC are separate circuits while the DSP processor and memory are on a single chip. In some cases, the RAM and ROM may also be in separate IC packages.

The music from a stereo receiver contains frequencies up to 20 kHz. What is the minimum sampling rate to ensure that all of the music information is retained in digital form?

   a. 10 kHz
   b. 20 kHz
   c. 40 kHz
   d. 55 kHz

---

16   (*c.* 40 kHz)   Initially, DSP systems were costly and complex. Over the years, DSP circuits have declined in price while becoming even more capable and inexpensive. Software has been developed to simplify the design of DSP circuits. Today, DSP is used in many modern electronic applications. The modem mentioned earlier is a good example. DSP is also used in TV sets, digital cellular telephones, CD players, personal computer sound circuits, and even in some toys and video games. As DSP tech-

(continued next page)

niques are perfected and as semiconductor technology permits smaller, faster, cheaper circuits, DSP will see increasing applications.

Here is a general summary of applications that benefit from DSP:

1. Telecommunications: Almost anything having to do with telephones, wired or wireless
2. Speech, voice, and music: Synthesis, recognition, enhancement
3. Graphics and imaging: 3D, image compression and decompression, enhancement, pattern recognition, machine/robotic vision
4. Consumer: DVD disk processing, new digital high-definition TV, hearing aids, radar detectors, toys and games
5. Industrial/control: Robotics, motor control, control of mechanical devices
6. Military: Cryptography, secure communications, radar and sonar processing, navigation

The leader in DSP technology in the United States and the world is semiconductor manufacturer Texas Instruments (TI). Their first DSP processors appeared in the mid-1980s. Today, their processors are used in more embedded DSP applications than any other processor. Other companies making DSP integrated circuits include Analog Devices, Lucent (formerly AT&T), and Motorola.

The TI TMS320XX series ICs are the most widely used DSP circuits. Let's take a more in-depth look at this popular embedded processor. The first device was the TMS320C10, while today a more recent version is the TMS320C80.

Figure 7-10 shows a general block diagram of the TMS320C50. It is one of several models of the TMS329C5X family of DSP with similar features. The X at the end of the model number designates one of 24 different models. The C50 model features Harvard architecture with separate but on-chip program and data memories. The program ROM holds up to 2K of 16-bit instruction words. The data RAM holds up to 512 16-bit words. An additional 9K+ × 16-bit RAM is available for more data, but may also contain a program. Special instructions allow the program/data memory to be configured as either program or data space. While most DSP programs are mathematically complex, they are relatively short, so only a small program ROM or RAM is needed.

The TMS320C50 contains a 32-bit ALU. It will add, subtract, multiply, and divide 32-bit fixed-point numbers. A key feature of all DSP computers is special hardware to perform high-speed binary multiplication. Most DSP algorithms reduce to a series of fast multiplications and additions. DSP ALUs are optimized for such operations. The C50 has one instruction that performs a 16 × 16 bit multiplication in addition to an add operation in one clock cycle.

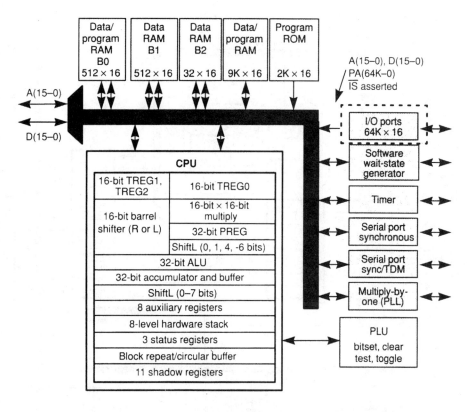

Fig. 7-10. General organization of the TMS320C50 DSP chip. *Courtesy Texas Instruments.*

A more complete diagram of the TMS320C50 is given in Figure 7-11. First, locate the program ROM, data RAM, and the additional data/program RAM. The program ROM is in the upper left side of the drawing. The data and data/program memories are at the bottom left of the figure. Next, identify the separate 16-bit program and data buses characteristic of Harvard architecture. An interesting feature of the TMS320C50 is the separate stack area located in the upper center of the diagram. The stack is RAM for eight 16-bit words to be stored during an interrupt or subroutine operation.

The main computing register is a 32-bit accumulator (ACCH/ACCL) located in the lower righthand area of the diagram. Associated with the ACC is the ALU and several shifters (SFL and SFR). These are high-speed logic circuits that shift the 32-bit words left or right to perform some types of multiplication or division operation, value scaling, or sign manipulation. The 16×16 multiplier with its 32-bit product output is shown above the ALU. Note throughout the diagram the blocks that are labeled MUX. These are multiplexer circuits that permit two or more inputs to be routed to a single destination under program control.

The parallel logic unit (PLU) is used to perform logical operations on the data without affecting the accumulator

(continued next page)

functional block diagram

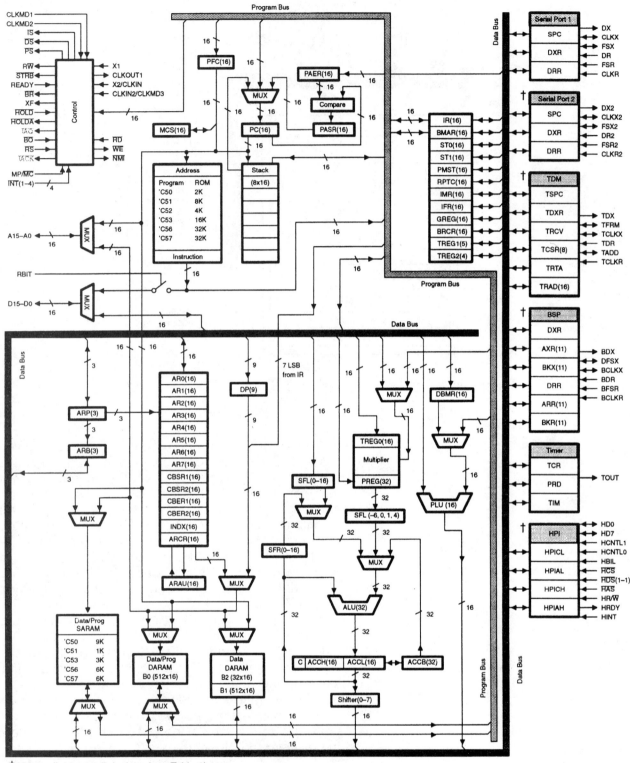

† Not available on all devices (see Table 1).

NOTES: A. Signals in shaded text are not available on
100-pin QFP packages.

B. Symbol descriptions appear in Table 3.

Fig. 7-11. Detailed diagram of the TMS320C50 DSP. *Courtesy Texas Instruments.*

content. It provides for high-speed bit manipulations and decision-making operations.

The eight auxiliary registers labeled AR0 through AR7 can be used to hold temporary data during calculations, but they are also used in indirect and indexed addressing operations. The addressing options also include the usual direct and immediate methods for accessing data and indexing.

The registers in the upper right portion of the diagram include the 16-bit instruction register (IR) and a group of registers that are used for status monitoring, block moves, and several different temporary operations.

The I/O circuits are located along the righthand side of Figure 7-11. The C50 has two serial I/O ports. These are full duplex synchronous serial ports for communicating with peripheral devices or other DSP units. Serial control register (SPC) is used to set up the serial port for the desired configuration, while the DRR and DRX are the serial receive and transmit registers. The time division multiplexed (TDM) port is a serial port that can communicate with up to seven other DSP units in larger, more complex systems.

On some models of the C5X the boundary scan port (BSP) and the host processor interface (HPI) are provided. The BSP provides a way to connect the processor to external test equipment to test its function and operation. The HPI allows 8- and 16-bit data transfers between multiple DSP units where the current processor serves as the host in a multiprocessing system. The C50 does not have these functions.

Finally, the C50 does have a built-in timer. It includes a 16-bit accumulator register/counter and a 4-bit prescaler. A variety of timing, delay, and clocking operations can be programmed to trigger interrupts.

The instruction set for the TMS320C50 is typical of most microcontrollers except for the special instructions that optimize the device for DSP. Many instructions execute in one clock cycle as small as 20 nS. Processing speeds of 50 MIPS (millions of instructions per second) are possible. The specialized instructions include multiply, divide, multiply and add, square, block move, and shift/rotate instructions.

Finally, find the address and data buses on the center left of Figure 7-11. The 16-bit address bus is labeled with bits A0 through A15. The data bus is labeled D0 through D15. These buses permit interfacing to larger external program and data memories if needed. The data bus is also connected to the external ADC and DAC for I/O operations.

Other versions of the TMS320XX family are faster, have larger on-chip memories, and have more extensive instruction sets. Some models have floating-point math operations to handle a wider range of numbers.

What math operations are unique to DSP?

(continued next page)

a. subtraction
b. division
c. square root
d. multiply and add

---

17 (*d.* multiply and add)   Answer the Self-Test
Review Questions on the diskette before going on to the
next unit.

Software

Software Defined

1   Computer hardware is really only half of a computer system. The computer hardware and peripherals perform no useful work until they are given a program. A program is a sequence of computer instructions that tells the computer what to do. Instructions are commands or statements in coded form that a programmer uses to define the computer's operations. The computer instructions are listed and executed sequentially to perform some useful function.

All of the programs that a computer uses are collectively known as *software*. Software is a general term used to describe any single program or group of programs. Together,

(continued next page)

the software and the hardware form a complete computer system. One is not usable without the other.

Most personal computer software is supplied on 3-1/2" diskettes or CD-ROMs. This software is transferred to the computer's hard disk where it can be quickly accessed when needed. The hard disk is a computer mass storage peripheral that stores data in the form of magnetized spots on a spinning disk. When it is to be used, the software is transferred from the hard disk into the computer's RAM. In embedded microcontrollers, the software is stored in a ROM.

A program is

    a. software
    b. made up of a list of instructions
    c. instructions that make the computer do something useful
    d. all of the above

---

2   (*d.* all of the above)   Software is the real interface between the user and the computer. The peripheral devices such as the keyboard, mouse, video display, and printer allow the user to communicate with the computer. But the peripheral devices are only the physical implementation of this communication. Programs called device drivers make the peripherals work with the CPU. Software is the interface that bridges the connection between the user and the hardware. See Figure 8-1.

An example illustrates this concept more clearly. When you are talking to the computer, you are not talking to the hardware. Instead, you are communicating with a program that controls all of the hardware. Assume that the computer displays on the video monitor a list of programs that are stored on the hard disk. You then select the desired program by highlighting it with a mouse and clicking the mouse button. The desired program is then retrieved from the hard disk, moved into main memory, and executed. This communication between you and the computer took place because of a program. The computer was programmed to output the menu of programs to the video display. You communicated with the computer by selecting the desired program with the mouse.

When you communicate with a computer, you talk to:

    a. hardware
    b. software

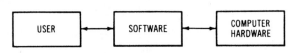

Fig. 8-1. Software is the real interface between the user and the computer.

---

3   (*b.* software)   The software used by a computer can be classified into two categories: applications software and systems software. All general-purpose computer systems use both types. *Applications software* are those programs employed by the user to perform some specific function. Some examples of applications software are an inventory-control program used for a small business, a special sequence of instructions that allows a microprocessor to control the displays in a gasoline pump, a program that plays a game of chess, a word processing or spreadsheet program, or a computer-aided design package for creating digital circuits. Although such application programs are sometimes developed by the user, more frequently publishers and software companies develop and sell application programs for specific computers. Our discussion here focuses on systems software.

Systems software consists of all the programs, languages, and documentation supplied by either the computer manufacturer or a software publisher. This is the software that facilitates the use of the computer and makes it more efficient and more convenient. Systems software helps manage the use of all other programs. Systems software is also applied by the user in developing applications programs of his own to perform some useful function. Systems software is also known as software-management or program-development software.

Figure 8-2 shows the hierarchy of systems software available for personal computers. The systems software is divided into the categories of languages and systems programs. Let's consider languages first. Go to Frame 4.

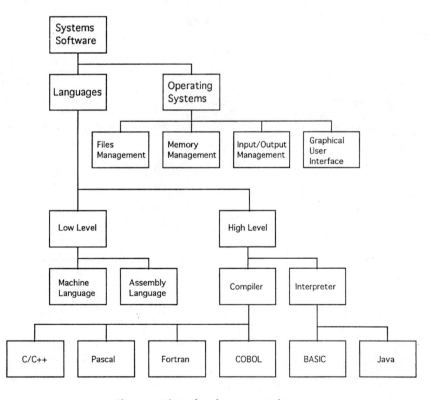

Fig. 8-2. Hierarchy of computer software.

(continued next page)

Computer Languages

4 A language is a system of communications that can take place between individuals or between an individual and a machine such as a computer. A language contains all of the symbols, characters, procedures, and syntax that an individual uses in communicating or exchanging ideas and information.

As you can see in Figure 8-2, there are two basic types of languages used in personal computers—*low-level languages* and *high-level languages*. These two levels are further divided into four categories—*machine* and *assembly languages* and *compiler* and *interpreter languages*. Machine and assembly languages are low-level languages, while interpreters and compilers are high-level languages. High-level languages are like our own English language. Low-level languages are more compatible with the hardware of the computer.

The lowest form of computer language is *machine language*. Machine language uses binary numbers to tell the computer what to do. Binary numbers use only the digits (bits) 0 and 1 to represent quantities and codes. Machine-language programming involves the use of the binary op codes, addresses, and data words to construct a program to perform some useful function. A simple binary machine-language program to perform the addition of two numbers (5 and 9) and store the sum (14) in memory location 00001000 (8) is shown to the right.

As you might expect, programming in binary is extremely time-consuming and error-prone. Machine-language programming can be simplified by using hex notation instead of binary. The previous binary machine-language program reduces to a somewhat simplified form in hex code, as shown in the table in the margin.

Whether you are using binary or hex notations, you are working directly with the computer hardware. You have full control over all its features. However, the disadvantage is that you must implement each and every minute detail of the program yourself. The sample program just given represents a significant amount of work to accomplish the simple task of adding two numbers. Yet in many applications this is desirable.

Machine language uses

    a. binary code
    b. hex values
    c. either *a* or *b*
    d. none of the above

Memory address	Content (instruction or address)	Meaning
00000000	00111110	Load accumulator with the next byte
00000001	00000101	Number to be loaded (5)
00000010	11000110	Add accumulator with the next byte
00000011	00001001	Number to be added (9)
00000100	00110010	Store accumulator
00000101	00001000	Least significant half of address
00000110	00000000	Most significant half of address
00000111	01110110	Halt
00001000	00001110	Sum (14)

Memory address	Content (instructions or addresses)
00	3E
01	05
02	C6
03	09
04	32
05	08
06	00
07	76
08	0E

5 (*c.* either *a* or *b*) Most computers are not programmed in machine language. Higher-level languages make programming much faster and easier. Assembly-language programming was developed to overcome the disadvantages of and objections to machine-language programming. Yet assembly-language programming has many of the same advantages as machine-language programming in that the programmer has direct access to and control over the registers, I/O ports, and other features of the computer. In addition, the disadvantages of having to work with binary or hex numbers are eliminated. Assembly-language programming overcomes this by allowing the programmer to use shorthand expressions for computer instructions, addresses, and data. Instructions are referred to by mnemonics, which are simply two, three, or four-letter abbreviations for the function performed by the instruction. For example, an instruction that tells you to load the accumulator may have the mnemonic LDA; an add instruction may have the mnemonic ADD; the instruction that tells you to store the accumulator might have the mnemonic STA; and so on. The mnemonic is simply a shorthand way of remembering an instruction. These mnemonics are used in place of the binary or hex op codes when writing a program.

In addition to using mnemonics to represent instructions, names can be used to refer to addresses or data. For example, you may use the name IOSUB to refer to the starting address of an input/output subroutine. The term CASH may refer to the amount of money one has on hand in his or her checking account. These names are used instead of the actual numbers. This allows the programmer to conveniently refer to them without knowing their exact values or where they are stored in memory.

Using these features, you can see that the machine-language program given earlier reduces to this simpler and easier to understand assembly-language program:

```
MVI     NUMB1
ADI     NUMB2
STA     SUM
```

The first instruction, MVI, loads one number to be added, NUMB 1 (5), into the accumulator. The next instruction, ADI, adds the second number, NUMB 2 (9), to the number in the accumulator. The sum (14) appears in the accumulator. The last instruction, STA, stores the sum in a memory location called SUM. The instructions MVI, ADI, and STA are the mnemonics. The names NUMB 1, NUMB 2, and SUM are symbolic memory addresses designating where the various data numbers are stored or will be stored.

Once you write an assembly-language program, you now have the problem of getting the computer to understand the program. The assembly language was written strictly to simplify the understanding and development of the program by the user. The user can readily learn the use of mnemonics for instructions and names for addresses and

(continued next page)

data, yet the computer cannot. All it understands is machine-language or binary numbers. How are the two made compatible? The answer is by the use of a special translation program called an *assembler*. An assembler is a program that is loaded into the computer's memory. Its primary function is to translate the user's assembly-language program using mnemonics and symbolic names into binary machine language. Just like a foreign-language interpreter will translate between two different languages, the assembler will convert the mnemonic program written by the user into assembly language into its binary machine-code equivalent.

In computer jargon, the assembly-language program prepared by the user is called the *source program*. The source program is typed in using a word processing program or a special program called an *editor*. The input to the assembler is the source program. The assembler then performs the translation and generates the equivalent machine code. This equivalent machine-code program is called the *object code*. The object code is the actual binary machine-code instructions and data that the computer can use. See Figure 8-3.

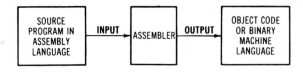

Fig. 8-3. Simplified illustration of how an assembler works.

Once the input source program has been translated into object machine code, the assembler is no longer needed. The newly generated object code can then be executed. In a typical computer system, the assembler is stored on the hard disk. When it is needed, the assembler is called up from the disk and loaded into memory. The source program is then entered by the user via the keyboard by way of the editor and then stored on a diskette or on the computer's hard disk. The assembler translates the source program into object code and stores it in another portion of the hard disk. The object code or program is then usually stored on a floppy disk. In this way, the program can be retrieved and executed again when needed. The object program can be loaded into RAM from the disk and executed or, in the case of embedded microcontrollers, the object code is burned into a PROM.

There are two important characteristics of assembly-language programming. First, an assembler is only good for one specific computer or microprocessor. An assembler for a Intel Pentium cannot be used for writing programs to be run on a Motorola Power PC. Second, assemblers typically translate a source program to an object program on a one-to-one basis. That is, one assembly-language mnemonic translates into one machine-code instruction or address. As you will see later, one statement in a higher-level language may translate to many machine-code instructions.

The output of an assembler is the

    a. source program
    b. object code

6 (b. object code) Higher-level languages were designed to make programming even easier. Some knowledge of the CPU and its architecture is required to program proficiently in assembly language. But with higher-level languages, no knowledge of computer operation or architecture is required. In fact, you do not even need to know binary numbers. Higher-level languages make it possible for almost anyone to program. Refer back to Figure 8-2.

You have already heard of some of the more popular higher-level languages such as BASIC, FORTRAN, C, C++, Pascal, COBOL, or Java. All of these languages are implemented as one of two different types of translators, compilers or interpreters. Like assemblers, compilers and interpreters are also translation programs. They take an input source program written in the higher-level language and convert it into the machine code that can be executed by the CPU.

When a programmer writes a program in compiler language, he or she does not write individual lines of instructions that correspond on a one-to-one basis with instructions in the computer's instruction set. Instead he writes expressions, statements, or commands that tell the computer what to do in a way unique to that particular language. These commands can be English-like statements or algebraic expressions. For example, the addition of two numbers A and B can be accomplished by the simple algebraic expression $X = A + B$. Recall earlier that it took two or more individual instructions to perform that same operation using machine- and assembly-language programming. The simple expression $X = A + B$ would be translated by the compiler or interpreter into the machine-code equivalent, which could be two, three, or more machine instructions.

Another typical higher-level language statement is PRINT "ANNUAL INVENTORY". This statement tells the computer to print the words "annual inventory". This English statement would be translated by the compiler or interpreter into many machine-code instructions that will perform that operation. You can see how much simpler programming becomes when you can tell the computer what to do by simply using English or math expressions.

Once a source program is written in a higher-level language using an editor or word processing program, it must be converted into machine code for the computer to execute it. This, of course, is done by a compiler or interpreter. The compiler, like the assembler, is a program that resides on the hard disk. When the compiler is needed, it is called up by the computer and loaded into RAM. The user retrieves the source program previously stored on a diskette or hard drive and puts it in RAM. The compiler then performs the translation. It translates the various statements

(continued next page)

and expressions into the machine code known as the *object program*. See Figure 8-4.

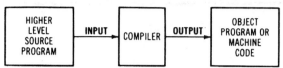

Fig. 8-4. The operation of a compiler.

Once the user's source program has been translated into object code, the compiler is no longer needed. The object code is then stored on a diskette. The object program can then be loaded into RAM and executed. The object code is stored in a PROM when an embedded microcontroller is used. Remember that the object program resulting from translation by a compiler (or an assembler) can be executed without the compiler being in memory.

The primary advantage of higher-level languages is that they greatly speed up and simplify the programming process. The program to solve a given problem can be written faster and easier in a higher-level language than in machine or assembly language. As you might expect, the development of higher-level languages has greatly expanded the use of computers. No longer must individuals be familiar with the hardware in order to write a program. All the programmer has to learn is a simple higher-level language that is much like his or her own language. Higher-level languages allow individuals with no computer knowledge whatsoever to take advantage of the many benefits that the computer provides.

The input to a compiler is called the

    a. source program
    b. object code
    c. editor
    d. assembler

---

7   (*a.* source program)   Another type of higher-level language is the interpreter. An interpreter is a program residing in computer memory that interprets and executes the higher-level language source program. Interpretative higher-level languages use statements, expressions, and commands similar to those used in compiler languages. The user preparing a program writes these expressions line by line using the unique syntax of that language to solve the problem. The interpreter is stored on the hard disk and then brought into memory (RAM) when it is needed. In some computers, the interpreter is in ROM. In this way, it is always available when needed. The main difference between the interpreter and compiler is that the interpreter must reside in memory along with the user's source program. See Figure 8-5.

The interpreter does not translate the user's source program into object code as compilers and assemblers do. Instead, the interpreter looks at each statement or expression and then performs the desired operations by selecting an appropriate subroutine within the interpreter itself that carries out the desired function. Although machine code eventually executes the program, the interpreter does not generate it. The machine-code subroutines used to perform

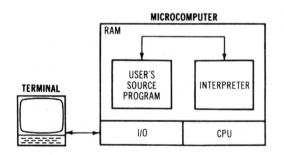

Fig. 8-5. The interpreter must reside in memory with the user's program.

specific operations already exist inside the interpreter. They are simply called upon as needed.

Interpreters are extremely handy and easy to use. Like compiler languages, they are easy to learn, and programs can be written quickly using the English and algebra-like statements and commands. The big disadvantage of an interpreter is its slow speed. Interpreters execute programs at a speed 10 to 20 times slower than equivalent machine code generated by an assembler or compiler. In some applications, this slow execution speed is not only undesirable but also unacceptable. However, keep in mind that computers execute programs at a high speed, and even with the inefficiency imposed by the interpreter they are more than fast enough for most applications.

Despite this disadvantage, interpreters are widely used. They are convenient and easy to use because they are totally interactive. The user can literally sit in front of the computer, converse with it, and solve problems directly through the interpreter. When a compiler language is used, first the source program must be entered and then compiled. The object code generated by the compiler is then loaded and executed. If the programmer wishes to change something in the program—adding features, correcting errors, or the like—the entire process must be repeated. The program changes are made and entered. The new source program is then recompiled. Finally, the new program is executed. With an interpreter, changes and additions can be made interactively. As soon as you discover the need to modify or add something, you can do it immediately through the interpreter. When using an interpreter, you will feel that you are truly speaking to the computer and commanding its attention. By far the most popular interpreter language is BASIC. Its popularity and ease of use is a result of its initial implementation as an interpreter.

An interpreter must reside in memory in order for the source program to be executed.

a. True
b. False

---

8 (a. True) Now let's take a look at some typical higher-level languages. One of the earliest higher-level languages was FORTRAN, developed in the late 1950s at IBM. FORTRAN means FORmula TRANslation, which gives a clue to its application. FORTRAN was developed primarily to speed up and simplify complex scientific and engineering calculations. It is a general-purpose, higher-level language that can be used in a wide variety of applications. However, it was primarily invented to optimize mathematical computation. Most higher-level languages are referred to as problem-oriented languages because they were developed to solve a particular class of problems.

(continued next page)

The primary language of FORTRAN is algebra. The rules and syntax of FORTRAN allow even the most complex mathematical formulas and expressions to be reduced to a simple, single-line statement. English-like statements or commands are further used to carry out specific operations. A typical example of a FORTRAN expression is given using the symbols in Table 8-1. The example given in the table shows how the well-known quadratic equation is solved using a FORTRAN statement.

FORTRAN is a compiler language. Once a programmer prepares the program, it must be entered into the computer, compiled into object code, and then executed. Despite its age, FORTRAN is still widely used in engineering and scientific applications. It is used primarily with supercomputers, mainframes, minicomputers, and engineering workstations.

Another widely used higher-level problem-oriented language is COBOL, which stands for COmmon Business Oriented Language. This language was developed in the late 1950s primarily to solve business data-processing problems, particularly those involving accounting, bookkeeping, inventory control, or various data storage and manipulation operations. Like all higher-level languages, COBOL uses English-like statements and algebraic expressions to state the problem. However, it is optimized for dealing with business-related problems. COBOL is also a compiler language that is used mainly on big business mainframes and minicomputers. Although it is no longer widely used for new programming projects, it is still a factor in large mainframe or legacy systems.

There has been a resurgence in the use of COBOL as large organizations reprogram their systems to deal with the "year 2000" (Y2K) problem. This is a serious potential problem because early programmers used only the last two digits of the year in programs. The program assumes that 57 means 1957. So, when the year changes to 2000, some computers will assume, use, and print a date of 1900. It is a major effort to reprogram computers so that errors and other problems do not disrupt business and government transactions.

Perhaps the single most widely used higher-level language is BASIC, which stands for Beginners All-purpose Symbolic Instruction Code. Like FORTRAN and COBOL, BASIC uses English-like statements and algebraic expressions to state the problem to be solved. BASIC was developed at Dartmouth College in the 1960s primarily to teach beginners the fundamentals of computer languages and programming. It was never intended to be much more than that. However, BASIC has been widely implemented both on large-scale computers and also on personal computers. It is also used in microcontrollers. Virtually every personal computer can use the BASIC language. BASIC was not optimized for any one particular class of applications. It has been used for solving mathematical problems, implementing games, doing engineering design, and general-purpose business data processing.

Table 8-1. FORTRAN Symbols for Mathematical Operations

Symbol	Operation
+	Addition
−	Subtraction
*	Multiplication
/	Division
**	Exponentiation

To solve the equation: $X = -B + \dfrac{\sqrt{B^2 - 4AC}}{2A}$

X = (−B + (B**2 − 4*A*C)**.5)/2*A)

Note: Raising an expression to the .5 power is the same as square root.

Although BASIC is primarily an interpreted language, BASIC compilers are also available. They are more efficient in their translation and are used instead of interpreters in demanding applications that require shorter programs that execute faster. Today the most popular version of BASIC is Visual BASIC, which combines the programming power of the language with a simplified way to create programs and applications that will run under the popular Windows operating system. BASIC is discussed in more detail later.

Which language is the most widely implemented on personal computers?

a. BASIC

b. COBOL

c. Fortran

ADA	This general language was developed in the 1980s by the Department of Defense as a universal standard language that could be used by all the military services to create software for computers and embedded applications such as navigation and weapons.
FORTH	FORTH was created by Charles Moore in the late 1960s and early 1970s. It is unusual in that it is both a compiler and an interpreter. A source program is first compiled then executed by the interpreter. This language is used in process control and other embedded applications.
Java	An object-oriented programming language created by Sun Microsystems in the early 1990s, this language was developed to implement programs that are to be used on the Internet. However, it is also very useful for creating other general-purpose programs. Java uses a unique software-virtual machine which interprets code written for it.
LISP	Lisp is short for LISt Processing. This language was created by John McCarthy at MIT back in the late 1950s and early 1960s. It is what is known as a symbolic processing language rather than a procedural language like most others. It is widely used in artificial intelligence (AI) program development. It is available either as an interpreter or a compiler.
Prolog	Prolog is short for PROgramming in LOGic. It solves problems using logical reasoning. Created in France, Prolog is widely used in the U.S., Europe, and Japan for artificial intelligence applications.
Smalltalk	This was one of the first languages created especially for object-oriented programming (OOP). It was developed at the Xerox Palo Alto Research Center in the late 1970s. Smalltalk is also an operating system that allows the user to customize the user interface.

Fig. 8-6. Other programming languages.

9 (a. BASIC) Many new languages were invented as the result of the introduction of personal computers in the 1970s. One of the most popular is Pascal. Developed by a Swiss professor named Niklaus Wirth, Pascal was designed to be easily taught and learned. Pascal is a powerful language that can be used to solve a wide variety of applications problems and to write higher-level languages and systems programs.

The main benefits of Pascal are its efficiency and portability. Its efficiency helps cut programming time and errors, so programmers become more productive with Pascal. Its portability makes programs easy to transfer from one PC to another. Pascal is available as a compiler only.

For many years, Pascal was one of the most widely used PC programming languages for creating new applications programs. But today the most widely used programming languages are C and C++. The C programming language was developed in the 1960s and 1970s at Bell labs as the language used to write the UNIX operating system. The C language was obscure for many years, but was rediscovered in the late 1980s. Since then it has become the language of choice for writing PC programs, having replaced Pascal in most applications.

C's virtue is that it is nearly as efficient as assembly language in creating fast, compact code. Yet, like other higher-level languages, it is easier to use. Like Pascal, C is designed for structured programming where software is created with modular pieces of code that are linked together. C is a compiler and often is supplied with libraries of reusable modules (subroutines) of code for math and other commonly used functions.

The C language has added more features to make it suitable for a new type of programming called object-oriented programming (OOP). Called C++, this newer version includes all the features of C with much more. Today, it is the

(continued next page)

single most popular programming language for personal computers and embedded controller applications.

Although C and C++ are the most widely used languages on PCs today, many others are available. Figure 8-6 lists some languages for special applications. Of these, the newest language, Java, bears special consideration.

Which language is *not* designed for object-oriented programming?

  a. Java
  b. Pascal
  c. C++
  d. Smalltalk

---

10 (*b.* Pascal)   Go to Frame 11.

Object-Oriented Programming and Java

---

11   Most programming languages are procedural languages, that is, they use what is generally known as the procedural method of programming. Procedural programming clearly separates the data to be processed from the procedures used to convert it from one form to another. Object-oriented programming combines the data with its various states or attributes with the program instructions or statements that specify the methods or procedures of processing. The three basic features of OOP that separate it from conventional procedural languages are encapsulation, inheritance, and polymorphism.

The basic element of an OOP language is an *object*. An object is computer code that combines data that may be numeric or symbolic and the methods or procedures defining operations that can be performed on the data. This combining of data and methods is called *encapsulation*. The objects are set up to be part of a *class,* a group of objects with similar characteristics.

Inheritance refers to a feature whereby objects may also inherit characteristics from other members of the same class. Data and methods in one object may be passed along to another object. You may create a new object that is based upon an existing object by using inheritance. The new object will add some new data and functions itself, but will draw upon those it inherited from another object.

Polymorphism lets you write a complete program that references many classes and objects rather than a single class. This allows you to achieve different results with a request from one object to another.

Programming in an OOP language is different from programming in a procedural language. With a conventional procedural language, you define the data by type and then write programs that tell how to process the data. In an OOP language, the programming process involves creating the objects and then sending messages between them. A *message* refers to one of the operations that can be performed on an object. Objects recognize messages they can perform. When an object receives a message, it carries out the operation on itself and generates a response.

The greatest benefit of OOP over conventional programs is modularity and reuse. Programs are easier to create because they are divided into many small modules that may be combined in a variety of ways. They can be used to create larger programs, or they may be reused by or interchanged with other programs. Because the major expense in creating software is programming time and cost, the creation of reusable code is an important benefit.

An object combines

    a. data and programs
    b. functions and classes
    c. data and methods
    d. procedures and inheritance

---

12   (*c.* data and methods)   Smalltalk was the first fully developed OOP language, and it is still used today. Other less popular OOP languages have also been developed. A major surge in the use of OOP occurred when the popular C language was modified to include OOP capabilities. The result was C++, currently the most widely used OOP language.

But a new language is quickly gaining popularity, Java. Java was developed by Sun Microsystems, a manufacturer of high-power engineering workstations and network servers. It was originally created to make the programming of Internet applications faster and easier. But in addition, Java has been "discovered" as a general-purpose programming language with some important benefits that no other language offers.

The actual syntax of Java is similar to that of C++, but numerous new features have been added to facilitate the creation of programs to be used in networks. Some key features of such programs are security and ease of use with large databases. Java programs are secure because Java will not pass along viruses and cannot read or write files on the user's system. But what makes Java really unique is that it can be used with any computer or operating system. The Java source code, which is standardized, is written once. Then, by use of a special interpreter program, this source code may be run on virtually any hardware, from a 30-year-old mainframe to a recent PC. This is accomplished by what is called the Java Virtual Machine (JVM).

In addition to defining its own syntax, Java also defines its own computer. This virtual computer is simulated in

(continued next page)

software. It has an instruction set like a microprocessor and an architecture. Figure 8-7 shows how Java works. You write the Java program in source code as you normally would any language with an editor or word processor. The source code is then compiled into machine code. In most other languages, the compiled object code is then loaded and executed. A traditional compiler translates the source code into machine code for a specific CPU such as an Intel Pentium or a Motorola Power PC. In Java, the compiler generates what is called Java byte code or Java virtual code. This is the code that is recognized and executed by the Java virtual machine. The Java virtual machine is an interpreter that is written for a specific CPU. It resides in the memory of the target computer. This interpreter then executes the Java byte code.

By using the JVM, the Java object code or byte code can be run on any computer that has a Java virtual machine interpreter installed. This interpreter is relatively short and easy to create for any computer, large or small, slow or fast. This means that the Java programs can be written once and then used by any computer regardless of its particular microprocessor or operating system. This makes the programs far more portable. The applications program can be written once and used many times with no further adaptation.

The primary disadvantage of Java is that it runs programs slower than other languages like C++. Because the Java Virtual Machine interprets the byte code developed by the compiler, execution speed is ten to twenty times longer than it takes for an equivalent C++ program. This disadvantage is overcome by using very fast processors. Because most programs are relatively short, execution speed is not usually an issue with today's super high-speed CPUs. Another solution is a special hardware Java machine rather than a software simulator. This is an integrated circuit that will run the Java code as fast or faster than a regular processor.

Java's original use was in creating short programs to be used with software, called *Web browsers*. A Web browser is a program that facilitates access to the World Wide Web, which is part of the Internet. A browser searches and provides the interfacing between a user and remote databases installed on special computers called *Web-site servers*. Such Java programs that run within the browser where the JVM resides are called *applets*.

Although Java's primary application will still be applets for browsers, it is quickly gaining acceptance as a generic programming language. Though it may never exceed C++ in popularity, it will nevertheless make great progress as a general-purpose language simply because so many new applications programs are associated with the Internet.

The Java Virtual Machine is a(n)

    a. compiler
    b. computer implemented in software
    c. interpreter
    d. special microprocessor that runs only Java code

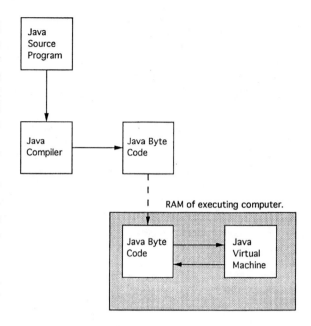

Fig. 8-7. How Java programs are written and executed.

13 (*b.* computer implemented in software) Go to Frame 14.

Operating Systems

14 *Operating systems* are a type of systems software used with all personal computers and some other types of microcomputer systems. They manage and control the computer's resources, thus simplifying the operation of the computer and the development of programs.

An operating system (OS) is a set of programs that collectively provide a way for the user to access and control all of the facilities of a computer. An OS automates the operation including virtually all of the hardware, peripherals, and applications software. This integrated collection of programs provides more efficient computer operation. Improved operating convenience is another benefit of an operating system.

Typically, operating systems are used only in larger microcomputer systems. For example, all personal computers, minicomputers, workstations, and servers need an OS. An OS is not really needed on most embedded microcontrollers. An OS requires some form of mass-storage media such as a magnetic hard drive, floppy diskette, or CD-ROM. Most microcomputer systems use operating systems that work primarily with disk mass storage. For that reason, these operating systems are referred to as *disk operating systems* (*DOS*).

Dedicated microcomputers built into a product require an OS.

> *a.* True
> *b.* False

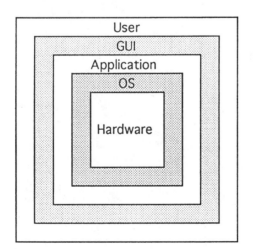

Fig. 8-8. The relationship between the user, the hardware, and the software in a computer.

15 (*b.* False) Figure 8-8 shows one way to visualize the operating system (OS). The OS effectively isolates the hardware from the user. The user communicates with the OS and the hardware through the applications programs and the graphical user interface (GUI). The OS fields all requests for service and totally eliminates the need for the user to deal with the hardware or related problems. The operating system allows the user to concentrate on his application and prevents him from having to spend a lot of time determining how things are to be specifically accomplished by the hardware.

An operating system has one major function: to manage the resources of the computer system. It allocates the resources on the basis of user need and systems capability. The four major functions of an operating system are:

(continued next page)

1. File and software management
2. Input/output and peripherals management
3. Memory management
4. CPU time management

The operating system software is usually stored on the hard disk in a personal computer. When the computer is first turned on, the operating-system program is transferred from the hard disk into RAM. With the operating-system program in memory, various operating-systems commands may be given by the user to perform a wide variety of operations, such as storing and retrieving programs and data files from the hard disk, a CD-ROM, or a diskette. Now let's consider the file and software management function.

The operating system is stored on:

    a. CD-ROM
    b. diskette
    c. hard disk
    d. tape drive

---

16   (*c.* hard disk)   Go to Frame 17.

Files and Software Management

---

17   The operating system supports a large library of typical user programs and files, stored on the hard drive. The user tells the operating system which program he or she wishes to use. The operating system retrieves the program from its source and brings it into main memory (RAM) for use. For software development applications, typical programs to be retrieved include an editor, an assembler, any of a variety of higher-level languages, a library of subroutines to perform often-used operations, and a debugger to help find problems in a new program being developed.

Files of data can also be created, stored, and retrieved under this system. A file may be a letter or proposal created with a word processor or a spreadsheet or a program written for a specific application.

Most operating systems have a hierarchy of storage capabilities. For instance, the storage area on a hard disk is usually divided into directories that contain subdirectories which, in turn, contain various files, programs, and data. This gives the user a way to organize his or her software and files. The OS manages the whole process of setting up and using these directories.

The hard disk is partitioned into:

    a. files
    b. directories
    c. records
    d. programs

18 (*b.* directories)   Go to Frame 19.

Input/Output and Peripherals Management

19   Input/output and peripherals management is another operating systems function. Operating systems also contain a variety of subroutines such as peripheral device handlers or drivers and interrupt-servicing routines. These programs run the video monitor, mouse, keyboard, printer, and other input/output devices. When writing new programs, most of the time and drudgery will be spent developing the input/output routines for talking to the various peripheral devices connected to the computer. Much of any given program concerns itself with inputting and outputting data. The operating system makes this easy by providing for all of these routines. When writing a program for a specific operating system, you simply need to specify what you wish to do with a given peripheral and the operating system will provide the desired code. This greatly simplifies and speeds up the programming.

A program that performs I/O operations for a specific peripheral unit is called a(n):

   a. device driver
   b. software controller
   c. I/O file
   d. language

20   (*a.* device driver)   Go to Frame 21.

Memory Management

21   Memory management is another function of the operating system. The programs in an operating system are capable of determining how much usable RAM a microcomputer has. The operating system also decides how this RAM is used. For example, it will determine where a system or applications program will be placed in RAM. The transferring or swapping of programs and data into and out of RAM takes place automatically under the control of the operating system. You don't have to concern yourself with this because the operating system does it automatically.

The operating system also manages the empty storage space available to it on the internal hard disk or on one or more external floppy disks. The operating system also decides where on the disk a user-created program will be

(continued next page)

stored. We say that such RAM and floppy-disk operations are "transparent" to the user.

The operating system can readily create and establish files for data input by the programmer or for programs generated. These files can be given a name and stored on the hard disk or a floppy disk. Again, the user does not have to know where these programs are stored. The program or data can be retrieved by giving the name to the operating system. The operating system will find that file and automatically bring it into RAM for the user.

A key feature of most OS memory-management software is *virtual memory*. Virtual memory is a software technique for making the CPU think that it has more RAM than it actually does. The OS allocates part of the hard disk drive to act as RAM. Although the CPU cannot actually fetch and execute a program directly from the hard disk, the OS can get that part of a program that won't fit into RAM and bring it into RAM a section at a time for execution. The virtual memory feature divides the program into pages and then swaps pages of a program between the hard disk and RAM. This allows the CPU to run a program that actually is too large to fit in RAM all at once. Most modern operating systems include virtual memory capability.

Virtual memory is:

    a. RAM the CPU only thinks it has
    b. the same as cache memory
    c. similar to ROM
    d. storage on the hard disk that appears to the CPU as extra RAM

---

22   (*d.* storage on the hard disk that appears to the CPU as extra RAM)   Go to Frame 23.

CPU Scheduling

---

23   Another common function of some operating systems is the scheduling of the CPU. The most efficient operation of any computer depends upon keeping the CPU busy. Because the CPU does all of the processing required of a computer, the CPU will generally be quite busy. However, there are times when inefficient, time-consuming I/O operations may be taking place and the CPU is idle. An operating system can detect such idle periods and give the CPU something to do. The operating system, therefore, schedules work for the CPU to keep it busy. This is particularly important in large computer systems that serve many people. Such systems may be operating in a time-sharing mode and have many inputs from a variety of users. To avoid inefficient idle periods, the CPU is sometimes given a backlog of work to do. The operating system then schedules the work to keep the CPU busy for maximum efficiency. Typical operating systems for microcomputers do not schedule CPU time since the system is basically dedicated to one user. Inefficiency does not necessarily matter for most microcomputer systems. If the user does not have work for the computer to do, it will simply sit idle.

Newer operating systems incorporate a feature called *multitasking*. Multitasking lets a user do two or more operations concurrently. An example of multitasking is printing out a letter while you continue to edit a spreadsheet. Another example is opening a graphics file, selecting a graphic, and then cutting and pasting it into a proposal you are writing with a word processing program that is already open and in use. Or, you could send a fax while working on a database update. Multitasking saves you time and makes you more efficient. Later versions of most OSs include some form of multitasking.

Doing two or more operations concurrently is called:

a. multiprogramming
b. multitasking
c. multicomputing
d. keeping several balls in the air at the same time

---

24 (*b.* multitasking)   Go to Frame 25.

Graphical User Interfaces

---

25   Although operating systems have been around for a long time, not until lately have they received what is called a *graphical user interface* (GUI, pronounced "gooey"). To use an early operating system, the user had to enter commands to tell the OS what to do. These commands were designated by mnemonics that were one, two, three, or four letters long. One example is DIR, which tells the OS to list the contents of its files and programs to the screen. The OS would then print out a directory of all of the files stored on the disk. Other examples are commands such as DEL for delete a file, COPY to copy a disk, and FORMAT to initialize a diskette for storage. Some OSs, such as UNIX, actually have a built-in programming capability. Although the commands have full control over the OS and the programming capability makes them very powerful, they are difficult for a casual user to employ. Using an OS is not simple and intuitive, so learning to use an OS requires extensive study and practice.

(continued next page)

In 1984, Apple Computer introduced the first graphical user interface for a personal computer operating system. Known as the Macintosh operating system, it featured a mouse, the pointing device that allows the user to move a pointer (the cursor) around on the screen. The mouse also had a pushbutton that could initiate selected operations by pointing to an item on the screen. This GUI used icons—small graphical symbols to represent programs, files, and applications—accessible on the computer's hard disk. Thus, you could easily select a program or file by pointing to the icon with the cursor and then choosing it by "clicking" on it with the mouse button.

In addition to icons, the GUI also incorporated *pull-down menus,* lists of operations that could be selected. Again the mouse selected the desired item and you initiated the operation by clicking. Most GUIs also use windows, small boxes presented on the screen that provide the user with options regarding what to do in a specific situation. Windows usually include buttons on which the user can click to make something happen. Obviously, navigating through a computer's capabilities with icons, menu choices, windows, and buttons makes using the computer faster, easier, and more convenient. Even beginning computer users with little or no training can quickly and easily select software and perform operations. This GUI changed the way that computers and software are used.

The Apple Macintosh operating system was quickly copied by Microsoft and adapted to other personal computers. Microsoft calls their GUI *Windows.* Windows, of course, is really a disk operating system with a graphical user interface. Now in its fourth and fifth generations, Windows is by far the most widely used operating system. Most other PC operating systems also use some form of GUI.

The operating system is probably the most powerful and important piece of software in a computer system. Along with the hardware, the operating system and other software form a complete, versatile, efficient, and easy-to-use system for the user.

Some of the most popular and widely used PC operating systems are Microsoft's Windows, IBM's OS/2 Warp, Apple's Macintosh OS, and the many varieties of UNIX, including versions from Novell, Digital, Sun, Hewlett Packard, and other vendors. Many versions of UNIX have a GUI known as X-Windows.

Which of the following is *not* usually part of a GUI?

    a. windows
    b. pull-down menus
    c. sound output
    d. buttons

---

26   (*c.* sound output)   Answer the Self-Test Review Questions on the enclosed diskette before going on to the next unit.

Programming

LEARNING OBJECTIVES

When you complete this unit, you will be able to:

1. Define the terms *programming* and *programmer.*

2. List and explain the five steps of the programming process.

3. Define the term *structured programming.*

4. Define the terms *algorithm, flow chart, pseudocode coding, testing and debugging,* and *documentation.*

5. Know the meaning of the terms *development system, cross assembler, cross compiler, simulator, emulator,* and *firmware.*

6. Explain the process of developing software for microprocessors.

---

1 *Programming* is the process of solving a problem with a computer. It is a multi-step process that involves identifying and defining the problem, developing a computer solution to it, and then preparing a sequence of computer instructions that solve the problem.

Most computer programming is done by individuals called *programmers.* Programmers are specially trained in problem definition and solution. They know how computers work and how to use the software to get a practical end result. Although most programming is still done by programmers, more and more nonexpert users are learning to program. As computers become easier to use, and as programming languages develop, this trend will continue.

Go to Frame 2.

The Programming Process

2   There are five steps in the programming process. These are:

1. Problem definition
2. Solution development
3. Program coding
4. Program testing and debugging
5. Documentation

The first step in programming is problem definition. Here you state as clearly as possible what you want to accomplish. Typically it is best to make a written statement of the problem. By actually writing out what is to be accomplished, you will think through the problem more thoroughly. Making charts, tables, graphs, drawings, or the like is also useful. Define your inputs, define the desired outputs, and then outline the processing steps required. Thorough problem definition produces a solution faster and easier. This is the most important step in programming.

Once the problem has been defined, you are ready to develop a solution. It is virtually impossible to give you guidelines on how to do this. Developing the solution is really the creative part of programming. The best way to learn it is by actually doing some programming. The more problems you solve, the more creative and proficient you will become. In any case, developing the solution is best done when the problem has been clearly defined and when you have a good working knowledge of the computer, its instruction set, and the programming language you are going to use.

The solution you develop will typically be a step-by-step procedure. Each step in the solution will cause some part of the problem to be solved. A step-by-step sequence of procedures for solving a specific problem is called an *algorithm*. An algorithm may be a mathematical formula or just a recipe-like list of steps.

An algorithm is a:

    a. recipe
    b. list of steps
    c. formula
    d. all of the above

---

3   (*d.* all of the above)   A useful tool in developing a solution to your programming problem and in preparing your algorithm is the *flow chart*. A flow chart is widely used by programmers to provide a visual or graphical presentation of the solution. Sometimes the solution is more easily seen or understood by blocking out a problem in graphical or visual form.

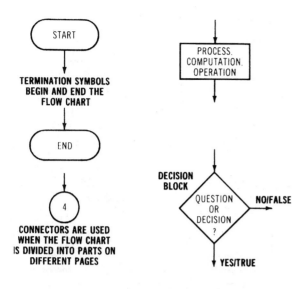

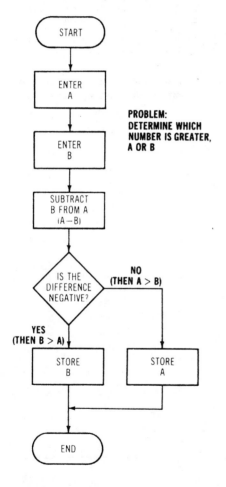

Fig. 9-1. Flow chart symbols.

PROBLEM:
DETERMINE WHICH
NUMBER IS GREATER,
A OR B

Fig. 9-2. A flow chart illustrating a computer program.

Flow charts are made up of a number of simple symbols that depict typical computer operations. Figure 9-1 shows the most commonly used symbols. The two most important symbols are the rectangle and the diamond. The rectangle simply specifies some particular operation. The diamond is a decision block. It usually asks a question that can be answered with a yes/no or true/false decision. This is the way most computers "think" or make decisions.

An example of a flow chart is given in Figure 9-2. This flow chart represents the solution to a typical computer problem. The problem in this example is to determine which of two numbers, A or B, is the larger. The algorithm is to subtract B from A. If the difference is positive, A is obviously greater than B. On the other hand, if the subtraction results in a negative difference, A is less than B, which means B is the larger. The flow chart illustrates the entire problem and solution.

Which flowcharting symbol indicates that a decision is to be made?

   a. circle
   b. diamond
   c. rectangle
   d. trapezoid

4   (*b.* diamond)   One of the best approaches to programming a solution is to use *structured programming.*

(continued next page)

Also known as *top-down design,* structured programming is the process of taking the problem and partitioning it into modules that perform the major functions of the solution. In structured programming, you break down a problem into smaller and smaller logical pieces. Each piece, or module, performs some distinct segment of the solution.

All of the modules are then linked by a hierarchical structure that shows their relationship. This structure looks like an organization chart for a company. An example is given in Figure 9-3. The main solution is divided up into three major modules. These modules are then divided into submodules that implement the primary module with smaller stand-alone programs. These submodules may then be further subdivided into even smaller logical segments. Some modules may be subroutines that may be used several times or called by any of the other modules or submodules.

Another feature of structured programming is to minimize or, if possible, completely eliminate unconditional branch operations. An *unconditional branch* or *jump* is a command or instruction that changes the order of a program from its normal, logical step-by-step sequence to something that might not be obvious. A common unconditional branch is the GOTO instruction. The tendency is to end one module with a GOTO so that the program branches to a submodule or some other logical segment of code. If too many GOTOs are used, the result is what is normally called spaghetti code. The result is a tangle sequence that is hard to follow in debugging or in maintaining the program.

The benefit of structured programming is that by dividing the program into smaller logical segments, it is easier to code, test, debug, and maintain. Also, it is much easier to work with a smaller program when devising a solution. So if you take the time to do the partitioning up front when designing the program, the actual development time will be shorter, and chances are that the program will work right the first time.

Which of the following is *not* one of the principles of structured programming?

  a. Minimize the use of GOTOs.
  b. Divide the solution into modules.
  c. Create a hierarchy of modules.
  d. Program simple functions first and then assemble into larger segments.

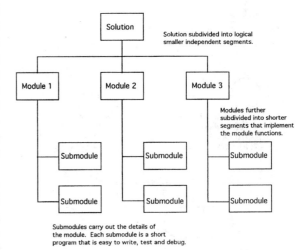

Fig. 9-3. Structured programming divides a problem into smaller and smaller submodules.

5 (*d.* Program simple functions first and then assemble into larger segments.) The next step in the programming process is *coding.* This is the process of converting your solution, algorithm, or flow chart into an actual computer program. Here you take the instructions, expressions, or statements of the language you are using and write them in a sequence that implements your algorithm.

A common procedure with many programmers is to first code the solution in pseudocode. *Pseudocode* is a hypothetical or made-up language that is some hybrid of an English language statement of the operation and the actual instruction, statement, or command of the programming language. It enables the programmer to write the sequence of operations on paper first before actually using the language itself. Often the pseudocode can be written from the flow chart. This procedure helps the programmer better see the logic of the procedure before committing to the final program code.

Programming involves much more than just the coding process, which most people associate with the term. Actually, coding usually takes 20 percent or less of the total time involved in programming. Studies have shown that 40 percent of a programmer's time is usually devoted to problem definition, structure design, and the development or identification of an algorithm.

Most programmers agree that coding is actually the easiest part of programming. Once you become familiar with the instruction set of the computer, the characteristics of the language you are using, and some basic programming tricks, coding becomes fast and easy. The use of subroutines greatly shortens and simplifies the coding process. On the other hand, defining the problem and developing a workable solution is much more difficult.

Coding means:

    a. finding an algorithm
    b. writing the program with a language
    c. defining the structure of the problem
    d. encrypting your data

---

6   (*b.* writing the program with a language)   Once the coding is complete, you will enter the program into the computer and test it. If you use assembly language or a higher-level language, you enter the program with the editor or a word processing program to create your source code. The computer will then assemble or compile your program into machine code. In either case, the result is the binary object code stored in memory. At this point you are ready for the next programming step, testing and debugging.

Testing involves verifying that your program works. You will want to "wring it out" well to be sure that it does what it is supposed to do. Invariably, computer programs do not work properly the first time. You may have made logical errors in sequencing, or perhaps you made errors in entering the code. If you used structured programming, you can test each submodule or subroutine by itself before linking together all of the various modules to perform the total operation.

(continued next page)

Whatever the problems, testing leads to *debugging*. Debugging is the process of working with the computer program and attempting to find errors and glitches. Most languages and program development software have built-in features that facilitate the debugging process. Once all the errors have been found and corrected, the program is ready to use.

The final step in the programming process is called *documentation*. This is an often-neglected but extremely important part of programming. Once the program is written and debugged, it is ready to use. The temptation then is to put the program into use and drop the project. However, it is extremely important that you document the program. Your problem definitions and solution work should be written up and filed away. Any charts, graphs, drawings, flow charts, and other materials that you developed should be saved. All program listings should also be retained. In fact, it is desirable to prepare a detailed, written statement explaining the algorithm and the procedures involved. This is important for program maintenance, because it may be necessary to change the program at a later date. You may wish to add a feature or correct some part of the program to accommodate new conditions. The original programmer may no longer be involved with the program. If a new programmer becomes responsible for changing a program, complete documentation is invaluable. Changing a program written by someone else is extremely difficult and, in some cases, virtually impossible without documentation.

The debugging and documentation part of the programming process represents about 40 percent of the programmer's overall time. The total allocation of a programmer's time is as follows:

Problem definition and solution	40%
Coding	20%
Debugging and documentation	40%
Total	100%

The major investment in any programming project is the programmer's time. A significant amount of time is usually required to develop a useful piece of software, which is why software is so expensive. In fact, the hardware is the least expensive part of the system in most embedded controllers.

One final word: Most programs will have to be changed over time. As the problem evolves, grows, or changes, the solution will have to be altered to meet current needs. New issues must be faced, or features may need to be improved to save time. In the case of a software product, improvements may have to be made to fix user-discovered bugs, improve performance, or keep the software competitive in the marketplace. It may be necessary to port the software to another CPU or operating system.

All of these conditions fall into the category of maintenance. In fact, more time is spent on maintenance than in creating the original software. And, as indicated earlier, the better the documentation the faster and easier the maintenance can be carried out.

Which usually takes the most time?

  a. maintenance
  b. coding
  c. debugging
  d. documentation

---

7   (a. maintenance)   Go to Frame 8.

Developing Software for Microprocessors

---

8   The programming process just described applies to any microcomputer, be it a general-purpose personal computer or an embedded microcontroller. However, the procedure is somewhat different if you are developing programs for microprocessors that are to be built into other equipment. Figure 9-4 shows a block diagram of the complete microprocessor software-development process.

One of the major applications of microprocessors is replacing conventional digital logic circuitry. The embedded controller or single-chip microcomputer must be programmed to perform the function that could be implemented with discrete logic elements. This program usually is stored in a ROM. Thus the microprocessor becomes dedicated to a specific application. The program in a ROM is typically referred to as *firmware*, that is, software that exists in an electronic component.

A microcontroller is simply an electronic component. It is useless by itself for developing the software that it needs. It needs a computer that will execute the instruction set of the desired microprocessor. This computer will then be used like a general-purpose computer to develop the software for the microprocessor. Today most microcontroller software development is done on a personal computer.

A significant amount of development software is available for personal computers. Typical programs include a text editor, cross assembler, cross compiler, simulator or emulator, and debugger. You normally use the text editor to write your program. Then you use a cross assembler or cross compiler to actually generate the machine code of the microprocessor.

A *cross assembler* is a system development program that runs on one computer for the purpose of translating assembly-language source programs into the machine code of

(continued next page)

the desired microprocessor. The cross assembler works like other assemblers in that it accepts mnemonic instructions with symbolic names for data and addresses. The mnemonics are those of the specific microprocessor to be used, not the mnemonics of the personal computer that is the host of the development software.

Once the assembly-language source program is written using the text editor, the cross assembler generates the desired machine or object code. Remember that the object code generated by a cross assembler cannot be executed by the personal computer itself because the instruction set for the computer is usually different from that of the microprocessor.

Cross compilers are also available on personal computers to produce programs for microprocessors. Cross compilers are used with higher-level languages. For example, the C and C++ languages are commonly used to write programs for microprocessors, as is BASIC. Of course, it is usually faster and easier to generate a program by using a higher-level or problem-oriented language, thus preventing you from having to think in terms of the specific instructions of the microprocessor when using the higher-level language. You can concentrate your efforts on problem-solving rather than on the coding process. Once you have written your program in the higher-level language with the text editor, the cross compiler is called upon to translate that program into the machine code for the target microprocessor.

Once the program has been written and the machine code generated, it is desirable to test the program. As indicated earlier, the time-shared computer is not capable of executing the microprocessor instruction codes. However, these computers usually have an emulator program available that will allow the personal computer to test the program. An *emulator program* is a system-level program that duplicates the execution of the microprocessor instructions on the personal computer. The emulator mimics the microprocessor so that the object code developed under the cross assembler or cross compiler can be tested. This allows the programmer to do any initial debugging. The problem with this procedure, however, is that the emulator sometimes does not operate in real time. It will not precisely duplicate the machine-cycle instruction execution rates of the microprocessor. The program will not be executed in exactly the same time frame as with the microprocessor. If time-related functions are important, the program should be thoroughly tested on a prototype of the microprocessor unit before being finalized.

A program that executes on a personal computer but mimics a specific microcontroller is called a(n):

    a. cross assembler
    b. cross compiler
    c. emulator
    d. editor

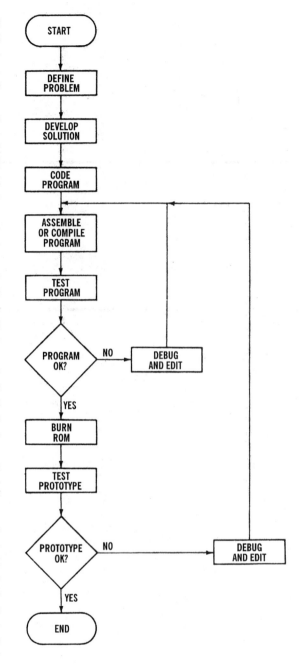

Fig. 9-4. Block diagram of the complete microprocessor software-development process.

9 (*c.* emulator)   A feature of some development systems is an emulator program that runs on the PC but also generates the signals produced by the target microprocessor. Using the input and output ports on the PC, the emulator can simulate precisely all input and output signals. A cable from the PC terminates in a connector that matches the pin configuration of the microprocessor IC package. This is plugged into the socket where the actual microprocessor is to be installed. The PC emulator completely simulates the microprocessor so that the software can be tested and debugged.

Also available for some of the most popular microcontrollers is a hardware emulator containing a target microprocessor with RAM instead of ROM. The PC downloads the object code into the emulator RAM. The emulator is then plugged into the socket to be occupied by the microcontroller.

Either of these emulation approaches lets you test a prototype of the equipment without having to reprogram the internal microcontroller PROM. Errors can be corrected and changes made in the PC or emulator RAM quickly and easily.

Finally, most development software includes a debugger, a program that helps to identify and isolate coding errors, logical errors, and other problems.

Regardless of the method used to produce the microprocessor software, the object code must be tested once it is developed. Typically this is done by "burning" the object code into ROM with a PROM programmer. The PROM programmer is a device that transfers a binary-coded program into a programmable read-only memory (PROM). The most commonly used PROM is the EPROM, erasable programmable read-only memory. The ROM may be internal to the microcontroller, or be a separate chip in some applications.

EPROMs are normally used when working with microprocessor breadboard prototypes. The EPROM may be part of a special version of the microcontroller chip designed especially for development of applications for which changes may have to be made. It could be a separate EPROM chip connected to the microcontroller. This is because sometimes the code must be changed to fix a programming error, adjust the timing, or add a desired feature. The EPROM is erased by applying ultraviolet light to it. EEPROMs are erased with an input pulse. Once the changes have been made in the development system and the new object code developed, the PROM can be reprogrammed. Although EPROMs can be used in the final production unit, they may be too expensive. Typically the final proven and tested code is sent to the microcomputer manufacturer to a ROM manufacturer where large quanti-

(continued next page)

ties of masked ROMs are made for the final production units.

Now answer the Self-Test Review Questions on the accompanying diskette before going on to the next unit.

Machine and Assembly Language Programming

1   Although programming is a subject best learned by actual practice, you have to start somewhere. So in this unit, you will follow the process involved in programming. The vehicle for doing this is two sample programs. We will follow the programing procedure outlined earlier and take you through the development of a program from problem definition to final code. The examples will provide you with the insight and the approach you will need to do your own programming.

First, we illustrate machine-language programming by using the individual instructions for the Motorola 68HC11 microcontroller. The 68HC11 CPU was discussed in Unit 7. As the program develops, we will introduce and explain a number of the 68HC11 instructions and addressing modes. Each instruction will be described as it is used in the program.

Next, we will show you the assembly language version of the sample programs. Assembly-language programming

(continued next page)

is similar, the only difference being that symbolic names are used for addresses and data rather than actual numbers when programming in assembly language.

Before you begin, you may wish to review briefly the architecture of the 68HC11 in Unit 7.

Now go to Frame 2.

Example Program 1

2 The programming process begins by defining and stating the problem. It is always best to make a written statement of the problem and provide as much information as possible about desired inputs and outputs. The problem to be solved by this example is as follows:

Write a program that converts two ASCII codes into their pure BCD equivalents. Then multiply the two BCD numbers and store product in pure binary form in a given memory location. The two ASCII numbers are initially stored in two designated-memory locations as the result of some I/O operation. Once these numbers are converted into BCD form, they will be replaced by their BCD equivalents. These two BCD numbers become the multiplier and multiplicand. The two numbers are then multiplied and the product is stored in a given memory location.

If you reread the problem carefully, you will find that it can be divided into two parts. The first part involves converting two ASCII numbers into their BCD equivalents. The second part involves multiplying the two resulting numbers. Always try to divide any large complex problem into many smaller, simpler procedures and then link them all together in the correct sequence to solve the problem. The ASCII to BCD conversion will occupy most of our programming effort because the 68HC11 has a multiply instruction to handle the second part of the problem.

The next step in the programming process is to develop a solution. You can find a clue to an algorithm for the ASCII to BCD conversion process if you look at the ASCII code table on page 178. Here you will find the hex equivalents for the numbers 0 through 9. If you convert these ASCII numbers into their binary equivalents and study them, you will see that the last four bits (least significant) of each ASCII number are the same as the 4-bit BCD equivalent. The four most-significant bits of the ASCII equivalent are not needed. Therefore, if we can eliminate the four most-significant bits, the remaining number will be the BCD equivalent. An example of this is as follows:

ASCII Code for 3	BCD equivalent
00110011	0011
First four bits not needed	

The ASCII binary number for the number 8 is

a. 1000
b. 10001000
c. 00111000
d. 11001000

---

3 (c. 00111000) One method of eliminating unwanted bits in a word is masking them by using the logical AND instruction. The AND function is widely used to perform a variety of bit manipulations in microprocessors. To remove an undesirable number of bits, you AND the ASCII word with a mask word carefully selected to retain the desired bits and get rid of those not wanted. The mask word is chosen such that binary 1's are placed in the equivalent positions of those bits you wish to save in the ASCII word. Binary 0's are placed in the bit positions of the mask word corresponding to those bits in the ASCII word you wish to remove. See the following example.

ASCII word	00111000	Last four bits will be saved
Mask word	00001111	First four bits will be masked out
Result after AND	00001000	

When you perform the AND operation, only those bits in the positions designated by the binary 1's are retained. The other bits are eliminated. To perform the AND operation, you simply execute the AND instruction.

In the binary number 10100110, we wish to retain the middle four bits and eliminate the two rightmost and two leftmost bits of the word. The mask word to accomplish this is _____. After the AND operation is performed, the remaining word will be _____.

a. 00111100, 00100100
b. 11000011, 11011011
c. 00111100, 01101001
d. 00001111, 00100100

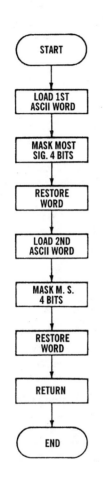

Fig. 10-1. Flow chart of ASCII-to-BCD conversion process.

4 (a. 00111100, 00100100) Using this algorithm, you can now develop a program that converts the ASCII numbers into BCD. You need to retrieve the ASCII numbers from their locations in memory, AND them with the appropriate mask word, and restore the BCD equivalents in the same memory locations. A flow chart illustrating this algorithm is shown in Figure 10-1. The program to implement it is given in Table 10-1. An explanation of the relevant 68HC11 instructions is given in Table 10-2. Before we explain the program step-by-step, it is important for you to

(continued next page)

understand the format. Each line in the program in Table 10-1 represents one byte of memory. In each location is an instruction op code, address, or data word to be used in the program. The first or leftmost column is the memory address given in hex. The second column contains the content of that memory location. If the content of the memory location is the op code of an instruction, the mnemonic for that instruction is given with its hex value in parentheses. If the word is an address or data word, the content is given in hex. The right-hand column is for comments and explanations. Each byte is described so that you will know what it is. Remember that Motorola uses a $ sign ahead of a number to designate that it is a hex number.

Table 10-1. ASCII-to-BCD Conversion Program

Memory Address	Memory Content	Explanation
10	LDAA (96)	Load accumulator A
11	2A	Address of first ASCII character
12	ANDA (84)	AND immediate
13	0F	Mask word
14	STAA (97)	Store accumulator A
15	2A	Address of modified ASCII character
16	LDAA (96)	Load accumulator A
17	2B	Address of second ASCII character
18	ANDA (84)	AND immediate
19	0F	Mask word
1A	STAA	Store accumulator
1B	2B	Address of modified ASCII character
1C	RTS (39)	Return from subroutine
2A	39	ASCII 9
2B	36	ASCII 6
2C	XX	Product of multiplication

Table 10-2. 68HC11 Instructions

Instr. Mnemonic	Operation	Hex op code
1. ANDA	Logical AND immediate. AND contents of accumulator A with second byte of instruction.	84
2. JSR	Jump to subroutine. Unconditionally branches to the 16-bit address contained in the second and third bytes of the instruction.	9D
3. LDAA	Load accumulator with content of memory location whose hex address is in the second byte of the instruction.	96
4. MUL	Multiply word in accumulator A by word in accumulator B. Put product in D register (A register extended with B register).	3D
5. RTS	Return from subroutine.	39
6. STAA	Store contents of accumulator A in memory location whose address is in the second byte of the instruction.	97
7. LDS	Load 16-bit stack pointer register immediate with a 16-bit address defining the location of the stack in RAM. The second and third bytes of this instruction form the address of the stack.	8E

Looking at the program listing in Table 10-1, identify where the two ASCII data words are stored.

a. 00, 01
b. 2A, 2B
c. 0F, F0
d. 84, 39

---

5   (b. 2A, 2B)   Refer to Table 10-1. The program begins with the first instruction stored in memory location $0010. (Remember that the dollar sign ($) means a hex value in Motorola 68HC11 format, and we often drop the leading zeros and just refer to the address as 10.) The first instruction is the LDAA or load accumulator A instruction. The next byte, stored in location 11, contains the address 2A, which is the location of the first ASCII character. The LDAA instruction then tells the computer to go to memory location 2A and load the word there into accumulator register A.

After the LDAA instruction is executed, the hex content of the accumulator register will be

a. 96
b. 36
c. 2A
d. 39

---

6   (d. 39)   The hex number 39 or the ASCII equivalent of the decimal number 9 will be loaded into the A accumulator. Note that the direct addressing mode is being used. LDAA is a two-byte instruction, the first byte being the op code and the second byte defining an 8-bit address between hex 00 and FF, the first 256 bytes of RAM where the desired data is stored.

The next instruction to be executed is stored in locations 12 and 13. This is the ANDA instruction, which is an AND immediate. This AND instruction will mask the desired bits. In location 13 is the mask word itself. The mask word in this case is hex 0F or binary 00001111. The 1's in the last four positions mean that the four least-significant bits of the ASCII word will be retained while the four most-significant bits will be eliminated. When this instruction is executed, the content of accumulator register A will be logically ANDed to the mask word in location 13. The result of this AND operation will be stored back in the accumulator replacing the word previously there.

After the ANDA instruction is executed, the hex content of the accumulator will be:

a. 39
b. 09
c. 0F
d. 84

7   (*b.* 09)   After the ANDA instruction is executed, the four most-significant bits will be replaced by binary 0's. The four least-significant bits defining the binary number 1001 or the BCD equivalent of the number 9 will remain in the accumulator. Note that this instruction is an example of the immediate addressing mode. Instead of referencing the mask word with an address, the mask word itself is stored directly after the instruction. In fact, the mask word is said to be a part of the instruction.

Once the AND operation is performed, the ASCII-to-BCD conversion for that word is complete. We now wish to store the BCD equivalent back into the same location where the ASCII word was originally stored. To do this we execute the next instruction in sequence, which is the STAA instruction stored in location 14. STAA is the mnemonic for store accumulator A. The number in location 15 defines an 8-bit address word where the number in the accumulator is to be stored. This, of course, is the same address originally used to load the accumulator. You will store the content of the accumulator back into location 2A. After the STAA instruction is executed, memory location 2A will contain the hex number 09.

It is important to note at this point that even though the number in the accumulator (09) is now stored in memory location 2A, the number still remains in the accumulator. Storing the number does not erase it.

Looking at the remainder of the program beginning with the LDAA instruction in location 16, you will see that it practically duplicates the first three instructions in the program. The next three instructions effectively fetch the second ASCII number from 2B, AND it with a mask word to perform the ASCII-to-BCD conversion, and then restore it in the same memory location.

The LDAA instruction in location 16 loads the next ASCII word into the accumulator, replacing the number previously stored there. After the LDAA instruction in location 16 is executed, the accumulator will contain the hex number 36. Hex 36 is the ASCII equivalent of the decimal number 6. Next, the ANDA instruction in location 18 is executed. It uses the same mask word as before, 0F. Executing this instruction removes the four most-significant bits of the word in the accumulator, leaving 0's in their place. The remaining four least-significant bits are retained.

When the ANDA instruction in location 18 is executed, the number left in the accumulator is 06. This instruction converts the ASCII number into its BCD equivalent, 6. Next, this BCD number is stored back in the location previously occupied by the ASCII number 2B. This is done with the STAA instruction in location 1A. Once the STAA instruction is executed, the number in the accumulator, 06, will be stored in memory location 2B.

At this point in the program, both ASCII numbers have been converted into their BCD equivalents. Only one additional instruction in the program is to be executed. This is the RTS instruction in location 1C that means Return from Subroutine. Any sequence of instructions that ends with a return instruction is a subroutine. Remember that a subroutine is a sequence of instructions that performs some specific operation. This ASCII-to-BCD conversion routine may be referenced in the main program by executing a Jump to Subroutine (JSR) instruction. The JSR instruction puts the subroutine into operation. The RTS instruction takes you back to the main program.

To execute the subroutine, a _____ instruction is used in the main program.

      a. RTS
      b. JSR

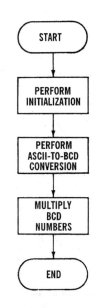

Fig. 10-2. Simplified flow chart of complete program.

8   (*b.* JSR)   Now let's discuss the part of the program that performs the multiplication. We also look at some other steps that are required to initialize the program and call the subroutine.

Figure 10-2 is a flow chart of the complete program. The first box in the flow chart is the initialization. There are some "housekeeping functions" that must be performed prior to execution of the program. One of these housekeeping functions is to designate the area in RAM where the stack will be located. This is done by setting the stack pointer to some specific address. Since we are going to use a subroutine in this program, an area of RAM must be set aside for the stack.

The next box in the flow chart of Figure 10-2 is the ASCII-to-BCD conversion subroutine. Finally, the actual multiplication of the two BCD numbers takes place. The program ends there.

The complete program is shown in Table 10-3. It is stored starting at location $00. The first instruction is LDS or load stack pointer. This is the program initialization referred to in Figure 10-2. The stack is assumed to be in location $00FF. The most significant part of the address (00) is in location 01, while the least significant part of the address (FF). The stack will contain the address to which the program will return after a subroutine.

Table 10-3. Complete Program

Memory Address	Memory Content	Explanation
00	LDS (BE)	Load stack pointer
01	00	Most significant half of address
02	FF	Least significant half of address
03	JSR (9D)	Jump to subroutine
04	10	Address of first instruction in subroutine
05	LDAA (96)	Load accumulator A
06	2A	Address of first number to be multiplied
07	LDAB (D6)	Load accumulator B
08	2B	Address of second number to be multiplied
09	MUL (3D)	Multiply contents of A and B
0A	STAB (D7)	Store product
0B	2C	Address of product

The next instruction, JSR, jumps to the ASCII-to-BCD subroutine stored beginning in location $10. The program counter is incremented to 05 and pushed on to the stack. The two ASCII characters are then converted to BCD by the subroutine. The RTS instruction at the end of the subroutine causes the content of the stack (05) to be retrieved and placed into the program counter. The next instruction to be fetched, therefore, is in 05.

The LDAA in 05 loads the A accumulator with the first BCD digit (9). The LDAB in 07 loads accumulator B with the second BCD digit (6). The MUL instruction multiplies the two values. The multiply operation assumes binary values, but in this case the binary values are equal to the BCD values. The result of the multiplication is $9 \times 6 = 54$, 00011110 binary or 1E hex.

The product of the multiplication appears in the D register, which is simply the designation of the A and B accumulators connected end-to-end to form a 16-bit register. The A accumulator is the most-significant half while the B accumulator is the least-significant half. In this case the hex content of the D register is $001E, which is our answer. We store the product 1E, which is in the least-significant half or the B accumulator. This is done by the STAB instruction in location 0A, which puts the product in location 2C. The program ends there.

This example illustrates a key point. Most programs are written by dividing the solution into subroutines and then calling them in sequence as needed. Other instructions are added to handle anything not covered by a subroutine.

Most programs are divided into logical segments, each performing some specific function. All of these segments are linked to produce the solution. Each of these segments is implemented as a(n)

    a. operating system
    b. language
    c. interrupt
    d. subroutine

---

9   (*d.* subroutine)   Although you can write a program in machine code like we did here, keeping track of all the hex op codes and addresses is aggravating. The best way to write this program is to use assembly language. No hex op codes are used, only instruction mnemonics and symbolic names for addresses and data values. The assembly language program for the previous problem is given in Table 10-4.

Table 10-4. Assembly language program for ASCII to BCD conversion and multiplication

	NAM	MPBCD
STACK	EQU	$00FF
ASC1	EQU	$002A
ASC2	EQU	$002B
PROD	EQU	$002C
MASK	EQU	$0F

	ORG	$0000
	LDS	STACK
	JSR	ASBCD
	LDAA	ASC1
	LDAB	ASC2
	MUL	
	STAB	PROD
	END	

	ORG	$0010
ASBCD	LDAA	ASC1
	ANDA	MASK
	STAA	ASC1
	LDAA	ASC2
	ANDA	MASK
	STAA	ASC2
	RTS	

You will recognize the same instruction mnemonics in this program, but there some new mnemonics as well. The program starts with the expression NAM MPBCD. NAM is one of several pseudo-ops used in assemblers. A pseudo-operation is one that only the assembler recognizes. In this case NAM tells the assembler to assign the name MPBCD to this entire file or program. You simply make up some name that relates to the program function. Here MPBCD means "multiply BCD" numbers.

The EQU pseudo-op is used in the next several lines to assign a symbolic name to the address locations where data is stored or to the data itself. Whenever the assembler encounters the term STACK in the program, it knows that the stack starts at $00FF. The term PROD for product is given to the memory location $002C where the product of the multiplication will be stored. ASCII number 1 (ASC1) is in location $002A, and that is assigned by the next EQU statement. The second ASCII number 2 (ASC2) is assigned to location $002B by the next EQU statement. Finally, the mask used to eliminate the first four bits of the ASCII number is assigned the value $0F by the last EQU statement.

The term used to describe an operation that the assembler, but not the microprocessor, will recognize and perform is called a

    a. quasi-op
    b. pseudo-op
    c. do-wop
    d. op code

10 *(b. pseudo-op)*   Continuing with the program in Table 10-4, note that the next pseudo-op encountered is the ORG statement. This means "origin" and is a way for you to tell the assembler that you would like the first instruction in the program to start in location $0000. The main program begins on the next line with the LDS instruction. Note the use of the symbolic name with each of the instructions instead of the hex address or data value.

The second part of the program in Table 10-4 is the subroutine that converts ASCII values to BCD. We name it by placing its symbolic name (ASBCD) before the beginning of the program. Now, to call the subroutine you write the jump-to-subroutine instruction, JSR, followed by the symbolic name of the subroutine. An ORG pseudo-op at the beginning of the subroutine tells the assembler where you want it to reside in memory. Otherwise the listing is the same as the machine-language version of the program given earlier except for the symbolic names of the addresses and data. The END pseudo-op tells the assembler that the program is finished.

You keyboard this program into a computer using an editor program, creating the source code. The assembler then translates it into the object code, which is the actual machine code that runs on the computer. The assembler eliminates the need for you to look up hex op codes to keep track of memory address locations. It will even figure out the branch instruction relative address value. Overall, the assembler saves you time and makes programming easier.

Which instruction in the previous program ends the subroutine and gives the operation back to the main program in Table 10-4?

      a. JSR
      b. END
      c. LDS
      d. RTS

---

11 *(d. RTS)*   Now let's take another programming example to illustrate I/O operations, loops and branches, and the use of the index register. Here is the problem:

Write a program that will transfer 128 bytes of parallel data to an external peripheral if a READY signal from the peripheral is binary 1. If the READY signal is binary 0, loop and wait for a binary 1. Use 68HC11 port B to output the data and port C to input the READY signal on pin PC7, the MSB. Port B's address is $1004; Port C's address is $1003. The 128 bytes of data are stored in memory locations $0020 through $003F (128 decimal=1F hex).

The flow chart for this program is shown in Figure 10-3. Figure 10-4 illustrates the hardware set-up. Refer to the

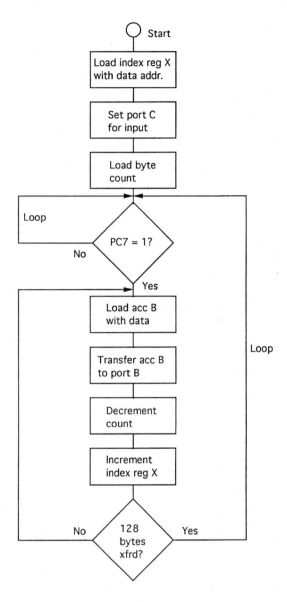

Fig. 10-3. Flowchart for I/O program.

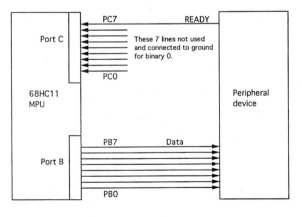

Fig. 10-4. Hardware configuration for I/O problem.

block diagram of the 68HC11 in Unit 7 if you need a refresher. The program listing is in Table 10-5. Table 10-6 explains additional 68HC11 instructions to be used. Refer to these as you read the following explanation.

Table 10-5. I/O Program

Address	Content	Explanation
00	LDX (CE)	Load index register X with address OF DATA $0020
01	00	
02	20	
03	CLRA (4F)	Clear accumulator A
04	STAA (B7)	Write 00 to DDRC for port C to make input
05	10	
06	07	
07	LDAA (86)	Load byte count immediate (128)
08	80	Hex value of 128 decimal
09	LDAB (D6)	Load accumulator B (Read port C)
0A	10	
0B	03	
0C	BPL 2A	Branch if plus, branch back 5 locations
0D	FB	
0E	LDAB, X (E6)	Load accumulator B with data at location In index register X
0F	00	
10	STAB (F7)	Write data byte to port B
11	10	
12	04	
13	DECA (4A)	Decrement count in accumulator A
14	INX (08)	Increment index register X
15	BNE (26)	Branch if not equal to zero, branch back 9 locations
16	F7	
17	BRA (20)	Loop back to 09 and repeat
18	F0	

Table 10-6. Additional 68HC11 Instructions

Instr. mnemonic	Operation	Hex op code
BNE	Branch always to address = PC + second byte value	26
BPL	Branch if plus (N flag = 0) to address = PC + second byte value	2A
BRA	Branch always to address = PC + second byte value	20
CLRA	Clear accumulator A to zero	4F
DECA	Decrement accumulator A (subtract 1 from A)	4A
INX	Increment index register X (add 1 to X)	08
LDX	Load index register X immediate with value in second and third bytes	CE

As you go over the program, keep in mind that the 68HC11 sets aside 64 memory locations starting at address $1000, which are actually 8-bit registers dedicated to I/O operations. Remember that the 68HC11 uses memory-mapped I/O where all I/O ports and other functions are given a memory address. These 64-address locations are on the chip.

(continued next page)

The first part of the program from memory locations 00 to 0C contains all of the set-up steps. Index register X is loaded (LDX) with the starting address of the data $0020. Accumulator A is cleared by the CLRA. All zeros in accumulator A are then stored in memory location $1007 by the STAA. This location is assigned to the data direction register (DDRC) for I/O port C. If all zeros appear in this register, port C is set up for input operations. Storing all 1's in the DDRC sets up port C for output operations. Any bit can be configured as an input or output by setting or resetting the bits in the DDRC.

Next we set up a counter that will keep track of how many bytes have been transferred. Accumulator A is loaded with the word count, in this case 128. This register will be decremented by one for each byte transferred.

The program begins by reading port C with the LDAB instruction. If the READY bit on PC7 is 0, bit 7 (MSB) in accumulator B will be 0. Note that all of the other input bits to port C are grounded, making them binary 0s. Bit 7 is also the sign bit of the accumulator. If bit 7 is 0, the value in accumulator B is +. If bit 7 is 1, the value in accumulator B is a 2's complement negative number. Our goal is to detect when a negative sign bit occurs, thereby indicating that the READY input line is 1.

We check this condition with the BPL instruction. This instruction looks at the negative bit (N) in the 68HC11 condition code register. This bit will be set if bit 7 in accumulator B (or A) is binary 1. The BPL instruction causes a branch if the N bit is 0, indicating a plus condition. In this case, the BPL branches back to the LDAB instruction, which again reads input port C looking for bit 7=1. The BPL again checks the N bit. The program loops by repeatedly executing the LDAB and BPL instructions until bit 7=1.

The BPL uses the relative addressing mode where the value in the second byte of the instruction is added to the program counter to form the address of the next instruction to be fetched. In this case, the second byte is FB or 11111011. Interpreting this as a 2's complement number, you can see its value. To find this value, perform a 2's complement conversion on FB hex by changing all 0's to 1's and all 1's to 0's, then adding 1 to the LSB.

Value	11111011
1's complement	00000100
Add 1	1
2's complement	00000101

This value is 5, and it is negative because bit 7 in the original value is 1.

As the 68HC11 fetches instructions, it increments the program counter (PC) so that it points to the next instruction to be executed. When the PC is at 0C, the BPL is fetched and the PC incremented twice to 0E. If a branch occurs, the second byte of the instruction is added to the PC. This gives:

0E	00001110
FB	+11111011
09	00001001

The BPL branches back to address 09 where the LDAB is executed, thus reading port C again. This continues until bit 7 = 1. The BPL condition (branch if plus) is *not* met, so no branch occurs. Instead, the next instruction in sequence is executed, LDAB, X in location 0E. Accumulator B is loaded with the data from the memory location formed by adding the second byte of the instruction (in this case 00) to the index register previously set to $0020, which is the starting address of the data.

The data word is then outputted to the peripheral by the STAB instruction to $1004 or port B. The count value in accumulator A is decremented by the DECA to 127, indicating that one byte was transferred and there are 127 left to go. The index register is incremented by one so that its address points to the second data byte in $0021. A BNE (branch if not equal to zero) instruction checks to see if the count is zero. The Z bit in the CCR is not 1 as it would be if accumulator A was zero, since the accumulator A has 127 in it. The BNE causes a branch back to 0E where the next byte is loaded into accumulator B and transferred to port B. The count is decremented and the index register is incremented. This loop continues until all 128 bytes have been transferred. When that occurs, the count in accumulator A is zero. The BNE does not branch this time. The next instruction in sequence, BRA, causes a loop back to location 09 where the READY bit is again tested. The program waits again until READY is 1. It then transfers another 128 bytes. It is assumed that some other program generates or captures the data that is put into locations $0020–$003F.

How many loops are there in this program?

    a. 0
    b. 1
    c. 2
    d. 3

---

12   (*c.* 2)   There are two loops, one formed by the BPL instruction as it monitors the READY input line and another that decrements a counter in accumulator A that tells how many bytes have been transferred. The BNE and BRA instructions form the loop.

Now try your hand at writing the assembler version of the program in Table 10-5. Just use the pseudo-ops as described earlier. Also, give a name or label to those instruc-

(continued next page)

tions to which the program branches if the condition of a branch instruction is met or not met.

Go to Frame 13.

---

13 The assembly program is given in Table 10-7.

Table 10-7. Assembly Language I/O Program

	NAM	IOPROG
PORTB	EQU	$1004
PORTC	EQU	$1003
DDRC	EQU	$1007
COUNT	EQU	$1F
DATA	EQU	$0020

	ORG	$0000
	LDX	DATA
	CLRA	
	STAA	DDRC
	LDAA	COUNT
RDY	LDAB	PORTC
	BPL	RDY
MORE	LDAB, X	DATA
	STAB	PORTB
	DECA	
	INX	
	BNE	MORE
	BRA	RDY

Please note the following:

1. RDY is the name or label given to the first LDAB instruction that reads port C. If the READY input is not 1, the BPL loops around and executes the LDAB again reading port C looking for a binary 1. It will simply execute this loop until the READY line is 1.
2. MORE (meaning there is more data to be transferred) is the label given to the second LDAB instruction, which is to be repeated for each data word to be outputted.

As a final exercise, you should try to write a simple 68HC11 program. Here is the problem:

A switch is connected to port C input pin PC7 just like the READY output line of the peripheral device in Figure 10-4. If the switch is open, PC7 will be +5 volts or binary 1. If the switch is closed, PC7 is zero volts or binary 0. You want the 68HC11 to read the state of the input and turn on an LED connected to pin PB0 on port B. The LED will turn on if the switch is open and turn off if the switch is closed. Use any parts of the previous programs that might work for you and modify them as needed. Use any of the previously explained 68HC11 instructions given in Tables 10-2 and

10-6. Start the program in location $0000. Use either machine or assembly language formats as you wish.

Go to the next frame for one possible solution.

---

14 Here is one solution.

ONE	EQU	$01
ZERO	EQU	$00
PORTB	EQU	$1004
PORTC	EQU	$1003
DDRC	EQU	$00

	ORG	$0000
	STAA	DDRC
	LDAA	PORTC
	BNE	OUT
	LDAB	ZERO
	STAB	PORTB
OUT	LDAB	ONE
	STAB	PORTB
	BRA	

The program starts by storing 00 in the DDRC making port C an input port. The next instruction reads port C. If the switch is closed, the PC7 bit will be 0. The BNE instruction says to branch on negative. Since PC7 is zero, the MSB in accumulator A is 0 meaning positive. Therefore, no branch occurs. The next instruction in sequence is executed. This LDAB loads zero into accumulator B. The STAB outputs this to port B, which produces a 0 on PB0, turning the LED off. The BRA branches the program back to the beginning to monitor the switch input.

If the switch is open, PC7 is 1, therefore the MSB in accumulator A is 1, meaning a negative value. The BNE detects this and branches to location OUT. At location OUT is a LDAB that loads 01 into accumulator B making the LSB 1. This is outputted to port B with the STAB instruction. The LED turns on. The BRA loops the program back to the beginning to again monitor the switch status.

Note: This program assumes that the switch does not produce contact bounce which will cause multiple 0 to 1 transitions on PC7. In pracctice, a delay subroutine is used to make sure the bounce has ended before recognizing the state of PC7.

Answer the Self-Test Review Questions on the enclosed diskette before going on to the next unit.

American Standard Code for Information Interchange

MSB						LSB
6	5	4	3	2	1	0

	Bits 6, 5, 4						
Bits 3, 2, 1, 0	000	001	010	011	100	101	110	111
---	---	---	---	---	---	---	---	---
0000	NUL	DLE	SP	0	@	P	'	p
0001	SOH	DC1	!	1	A	Q	a	q
0010	STX	DC2	"	2	B	R	b	r
0011	ETX	DC3	#	3	C	S	c	s
0100	EOT	DC4	$	4	D	T	d	t
0101	ENQ	NAK	%	5	E	U	e	u
0110	ACK	SYN	&	6	F	V	f	v
0111	BEL	ETB	'	7	G	W	g	w
1000	BS	CAN	(	8	H	X	h	x
1001	HT	EM	)	9	I	Y	i	y
1010	LF	SUB	*	:	J	Z	j	z
1011	VT	ESC	+	;	K	[	k	{
1100	FF	FS	,	<	L	\	l	|
1101	CR	GS	-	=	M	]	m	}
1110	SO	RS	.	>	N	^	n	~
1111	SI	US	/	?	O	—	o	DEL

Programming in the BASIC Language

LEARNING OBJECTIVES

When you complete this unit you will be able to:

1. Define and use the following BASIC statements and commands: REM, LET, INPUT, PRINT, GOTO, GOSUB, RETURN, IF THEN, RUN, END, LIST, and CLEAR.

2. Read, understand, and interpret a simple BASIC program.

3. Write a simple BASIC program to solve a problem.

4. Define and use the BASIC statements FOR NEXT, READ, and DATA, and the functions RND, INT, and ABS.

5. Define the PEEK and POKE commands and explain how they are used.

6. Define the term *loop* and recognize loops in a BASIC program.

7. Define the term *strings* and use strings in a BASIC program.

8. Read, understand, and interpret a simple BASIC program using the preceding statements and concepts.

9. Write a simple basic program to solve a problem using the preceding statements and concepts.

10. Name two BASIC interpreters used in embedded microcomputers.

11. Interpret simple programs using BASIC-52 and PBASIC.

12. State the publisher of Visual BASIC and explain what it is and why it is beneficial.

Introduction to BASIC

1   One of the most widely used higher-level languages and perhaps the easiest to understand is BASIC (Beginners All-purpose Symbols Instruction Code). Like all higher-level languages, BASIC eliminates the need to learn or understand any specific computer or microprocessor architecture and instruction set. With such higher-level languages you can focus on the problem by having a flexible set of instructions, commands, and other tools that help you quickly develop a solution with minimum programming effort.

BASIC began life as a simple language designed to teach programming at Dartmouth College in the 1960s. In the 1970s it became the language of choice for personal computers. Today it is still widely used to write programs in all fields of business, industry, science, engineering, and education. It is even widely used to program microprocessors for embedded controller applications. BASIC is available in both compiler and interpreter versions. Interpreter versions are the most common. Because BASIC is so popular and widely implemented, we have chosen it to give you a feel for programming in a higher-level language.

Most BASICs are essentially comprised of similar types of rules, procedures, and syntax. BASIC on one computer is essentially the same as BASIC on another. However, most vendors of BASIC add to or otherwise enhance their version of the language and, as a result, there are many dialects of the language. Despite the different dialects, a program written in BASIC for one computer will run on another computer, assuming that the minor differences between the two dialects are resolved.

The dialect of BASIC that you will learn in this unit is essentially a generic subset of most BASICs. It is more than adequate to gain an understanding of how this language is used to solve problems. You can easily expand your knowledge of BASIC by referring to one of the many other books available on the subject or using an available version of the language.

As you have seen, problems are solved on a computer by writing a program. A *program* is a series of instructions that tell the computer what to do. In higher-level languages, however, you do not deal with individual computer instructions. Instead, you work with statements and commands. Statements are instructions that tell the computer specific things to do. Programs are made up of a sequential list of statements. Commands tell the computer what to do with a program. In BASIC these statements and commands are English words or abbreviations that give the computer a specific command or call for the solution of a specific problem. A list of common BASIC statements is given in Table 11-1. Please stop at this point and read the entire list before proceeding.

Table 11-1. BASIC Statements and Commands

Statement/ Command Form	Description
REM (text)	The remark (REM) is a nonexecutable statement used only for commentary on a program listing. Example: REM PROGRAM TO COMPUTE PAYROLL
LET Var = Exp	This statement assigns the value of the expression to the variable. Example: LET X = 4 Let Y = C – F
INPUT Var1 Var2, . . .	This statement allows you to read data from the keyboard and assign values to the variables. Example: INPUT 5 ?
PRINT "message"; Exp	The message or value of the expression is printed on the screen with both a leading and trailing space. Messages may be numbers or letters and are enclosed within quotations. Example: PRINT "HELLO" PRINT A + B – C PRINT COST; C*.15
GOTO Num	This statement causes an unconditional branch to the statement numbered Num and execution continues. Example: GOTO 150
GOSUB Num	The go-to-subroutine (GOSUB) statement causes an unconditional branch to the statement Num. When the RETURN instruction is encountered in the subroutine, program execution returns to the statement following the GOSUB. Example: GOSUB 750
RETURN	RETURN is always the last statement of a subroutine.
If Exp1 rel Exp2 THEN Stmt	This is a decision-making statement. If the test "Exp1 rel Exp2" is true, the statement after the "THEN" is executed. This statement can be any BASIC statement. The "THEN Stmt" part can be replaced by GOTO Num BASIC recognizes the relational (rel) operators: = < > < = > = Example: IF X = 9 GOTO 70 IF A > 6 THEN 50
RUN	This command starts the program at the statement with the lowest statement number.
END	When the interpreter encounters an END statement in the program, it stops program execution.
LIST	The LIST instruction writes the entire program to the screen.
CLEAR	The interpreter removes all program statements from the buffer when it encounters a CLEAR instruction. Variables are not cleared.

Vocabulary
1. Exp   Expression
2. Num   Number
3. Rel   Relational, relationship
4. Stmt   .Statement
5. Var   Variable

(continued next page)

A BASIC program is made up of a sequence of:

    a. instructions
    b. algorithms
    c. statements and commands
    d. objects

---

2   (*c.* statements and commands)   Many BASIC problems are solved by evaluating expressions. An expression is typically an algebraic presentation of some mathematical relationship. Several typical algebraic expressions are as follows:

$$Y = X^2 + 2X + 4$$
$$C = \sqrt{(A^2 + B^2)}$$
$$A = BH/.5$$

BASIC solves the problem by evaluating each expression and presenting the answer.

Expressions are made up of constants, variables, and operators. Constants are simply fixed values or numbers that are used in the expression. They may be integers (whole numbers) or decimal values, positive or negative. Some examples of constants are:

22.5	100.1101
–137	–.0068

Variables are unknown values or values that may change. Variables are usually represented in expressions by letters of the alphabet. In BASIC, any of the letters of the alphabet A through Z may be used to represent a variable. In fact, you can also use letters of the alphabet followed by a number to give a variable a name. Some typical valid variables are given below.

A	FR
C6	Z12

Operators refer to the arithmetic operations addition, subtraction, multiplication, division, and exponentiation (raising a number to a power). The operators used in BASIC are:

+ = addition	/ = division
– = subtraction	^ = exponentiation
* = multiplication	

Note that addition and subtraction are represented by plus and minus signs as in regular algebraic expressions. However, multiplication is designated by an asterisk and division by a slash. Exponentiation is represented by an arrow pointing upward. To perform arithmetic operations, these operators must be used. As an example, an expression is first given in the regular algebraic format and then expressed using BASIC operators.

Algebra	$X = [-B + \sqrt{(B^2 - 4AC)}]/2A$
BASIC	$X = (-B + (B^\wedge 2 - (4*A*C))^\wedge.5)/2*A$

Note that raising an expression to the .5 power is the same as square root.

What is the BASIC expression of this algebraic expression?

$$Y = (MX + B)/X^2$$

3 $(Y = (M*X + B)/X^\wedge2)$ Parentheses can also be used as part of the problem to distinguish the sequence of problem solution. If we write an expression without parentheses, the expression will be evaluated according to the rules of precedence, which say that any exponentiation is performed first, then any multiplication and division, and finally addition or subtraction. This can sometimes lead to confusion. Parentheses should be used where necessary to organize the problem so that it is solved in the correct sequence. By using parentheses, all confusion is erased. Expressions inside the parentheses are evaluated first and then the other operations are performed according to the rules of precedence. Although most versions of BASIC follow the above guidelines, some do not. Be sure to check the documentation for the BASIC you are using with regard to rules of order.

Before we go any further, let's take a look at a typical BASIC program. This program (Table 11-2) is simple because its only purpose is to compute the sum of two numbers, A and B. However, it will illustrate how BASIC statements and expressions are evaluated.

Table 11-2. Simple BASIC Program

```
10      REM EXAMPLE PROGRAM
20      PRINT "THIS PROGRAM COMPUTES THE SUM OF A AND B"
30      LET A = 7
40      LET B = 16
50      LET C = A + B
60      PRINT C
70      END
```

Each statement in the program is written on a separate line and begins with a line number. Usually the line numbers start with the number 10 and increase by 10. The computer executes each statement in line-number sequence. Statement numbers from 1 through 9999 can be accommodated, but it is customary to space each line

(continued next page)

number by a factor of ten. This allows you to insert line numbers later if you wish to change or add to the program.

Refer to Table 11-2. The first statement in line 10 is simply a remark (REM). The expression "EXAMPLE PROGRAM" is printed out only if the program is listed with a LIST statement.

The next statement, line 20, tells the computer to print the expression enclosed in quotation marks. The computer will print on the video screen virtually any message made up of letters and/or numbers if you simply put the expression in quotation marks after the PRINT statement.

To print the message BASIC IS EASY, you would write the expression _____.

---

4   (PRINT "BASIC IS EASY")   Lines 30 and 40 in Table 11-2 use the LET statement to assign values to variables. The statement LET A = 7 tells the computer that it must use the number 7 in the computation every time it encounters the letter A. The same is true of the LET B = 16 statement. Some BASICs allow you to assign variables by just typing A = 7 or B = 16.

The statement on line 50 shows the use of a complete but simple algebraic expression. Here C is defined to be the sum of A and B. LET C = A + B.

Line 60 says to PRINT C. This statement tells the computer to print the numerical value of C after it is evaluated by the LET statement on line 50. The last statement on line 70 is the END statement that tells the computer it is at the end of the program. To run this program, simply type the command RUN. Line numbers are not used with commands like RUN, CLEAR, and LIST. The computer immediately scans through the lines in sequence producing the following output.

RUN	(You type this)
23	(Computer responds with the answer.)

Of course, the sum of 7 and 16 is 23.

Now suppose that you want to review or modify your program. All you have to do is type LIST. The complete program as given in Table 11-2 will be printed on the screen. You can interactively change the program by inserting new statements of correcting errors.

Go to Frame 5.

5   Problem solving in BASIC is virtually identical to the problem-solving process with any computer language. As you recall, the first step of this procedure is to define the problem. Usually a written statement is best, but other information such as charts or graphs should be provided as needed. The available inputs and the desired outputs should be specifically stated.

Next comes the development of a solution. At this point you devise the procedure to use to solve the problem. The solution is usually expressed as a step-by-step process called an *algorithm.* The solution should also be expressed with a flow chart.

Finally, you are ready to program the solution. Writing the program is called *coding.* In the case of BASIC, coding refers to writing the statements, expressions, and commands that implement your algorithm. As you will find, coding in BASIC is considerably easier than coding in machine or assembly language. The computer gives you far more powerful capabilities in expressing and solving the problem.

Remember that once the program has been written, you will need to run it and test it. This often leads to the need for troubleshooting or debugging. Then you wrap up the entire process by creating a documentation package that describes the problem and its solution in detail.

The following example problem illustrates most of the steps reviewed above.

Problem Statement

Write an interactive BASIC program that will perform Fahrenheit-to-Celsius and Celsius-to-Fahrenheit temperature conversions. The user should be able to enter the temperature in either Fahrenheit or Celsius and the computer will compute the equivalent. The program should be self-instructional, meaning that it should state the purpose of the program and give the user instructions as needed.

Solution

The creative part of programming is developing the solution. There are usually several correct solutions. The one described here is typical. The keys to the solution for this problem are the well-known formulas for Fahrenheit/Celsius temperature conversion.

$$C = 5/9(F - 32)$$
$$F = 9C/5 + 32$$

(continued next page)

Add to this some instructions and the problem is solved. Write the two above expressions in legal BASIC format.

C = _____
F = _____

---

6 (C = 5/9*(F − 32), F = 9/5*C + 32)   Figure 11-1 shows a flow chart of the solution. Note particularly the decision symbol. You must ask the user if he or she wishes to convert F to C or C to F. Once the choice is made, the computation is easy. Also notice the INPUT statement. This statement lets you enter the temperature you want to convert.

Now refer to the completely coded program in Table 11-3 and go through it line by line. Lines 10 and 20 simply print a message telling what the program does. Line 30 is a PRINT statement with nothing following it. The result is a blank line that lets you space lines and data.

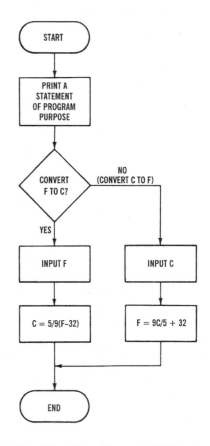

Fig. 11-1. Flow chart of example BASIC program.

Table 11-3. Complete Example BASIC Program

```
10      PRINT "THIS PROGRAM CONVERTS TEMPERATURE IN
        DEGREES FAHRENHEIT (F)"
20      PRINT "TO CELSIUS (C) OR CELSIUS (C) TO FAHRENHEIT
        (F)"
30      PRINT
40      LET Y = 1
50      LET N = 0
60      PRINT "TYPE 1 FOR F TO C OR TYPE 0 FOR C TO F"
70      INPUT A
80      IF A = Y THEN GO TO 100
90      IF A = N THEN GO TO 150
100     PRINT "WHAT IS THE TEMPERATURE IN F?"
110     INPUT F
120     LET C = 5/9*(F − 32)
130     PRINT "THE TEMPERATURE IN C IS"; C
140     GOTO 200
150     PRINT "WHAT IS THE TEMPERATURE IN C?"
160     INPUT C
170     LET F = 9/5*C + 32
180     PRINT "THE TEMPERATURE IN FAHRENHEIT IS"; F
200     END
```

Next the decision must be coded. This is done by assigning the value 1 to the answer yes (Y) and the value 0 to the answer no (N). The LET statements on lines 40 and 50 make this assignment. Yes means to do an F-to-C conversion. No means a C-to-F conversion. The PRINT statement on line 60 asks you to decide if you want to do a C-to-F or an F-to-C conversion. You are asked to respond by typing 1 or 0.

If the user types 0, what conversion is performed?

a. C-to-F
b. F-to-C

7 (*a.* C-to-F)   The INPUT A statement on line 70 allows you to enter your answer (A), 1 or 0. The computer prints a question mark (?) when the INPUT statement is executed. The computer then waits until you type in your response.

The statements on lines 80 and 90 analyze your answer. If the answer is 1, then A = Y and the program branches to line 100. If the answer is 0, then A = N and the program branches to line 150. This decision-making operation is done by the IF, THEN statement. If a stated condition is true, then the program jumps to a given line number to continue execution. If the stated condition is not true, the next line number in sequence is executed.

If the user inputs the answer 1, the program branches to line 100. If your response to the INPUT A statement is 0, the next statement executed is on line 90. The program then branches to line 150.

Now let us assume that you type 1 in response to the question asked in line 60 and the INPUT statement on line 70. The program branches to line 100. The PRINT statement there asks "WHAT IS THE TEMPERATURE IN F?" The INPUT statement on line 110 lets you enter the temperature in degrees Fahrenheit that you wish to convert to Celsius. When you type in the temperature and then press Enter, the next statement in sequence is executed.

This is the statement whose line number is:

> *a.* 90
> *b.* 100
> *c.* 120
> *d.* 150

8 (*c.* 120)   Using the temperature in degrees Fahrenheit that you entered, the LET statement computes the equivalent Celsius value and assigns it to the variable C. Next, the statement on line 130 causes the terminal to print "THE TEMPERATURE IN C IS" followed by the actual numerical value. Finally, the GOTO statement on line 140 causes an unconditional branch to line 200 where the END statement stops the program. A BASIC interpreter executes the program one statement at a time, in line number sequence.

Now, let's run through the entire program from the beginning as it would be executed on a typical personal computer.

First, to execute the program, type the command RUN and press Enter. When you do this, here is what happens on your computer. Note that the underlined numbers are the data you type in.

(continued next page)

```
THIS PROGRAM CONVERTS TEMPERATURE IN DE-
GREES FAHRENHEIT (F) TO CELSIUS (C) OR
CELSIUS (C) TO FAHRENHEIT (F).
TYPE 1 FOR F TO C OR TYPE 0 FOR C TO F
?1
WHAT IS THE TEMPERATURE IN F?
?34
THE TEMPERATURE IN CELSIUS IS 1.11111
END AT LINE 200
```

The program waits for you to respond at the two points where a question mark appears. This program tells you that a temperature of 34 degrees Fahrenheit is equivalent to a temperature of 1.11111 Celsius.

Now, you play the role of the computer. Execute the program yourself by writing on a separate sheet of paper the response of the computer if you select a 0 for C-to-F conversion, and if you enter a temperature of 37 degrees. Use the format just shown, and use a calculator to do the math.

---

9   (Your answer should appear as shown below.)

```
THIS PROGRAM CONVERTS TEMPERATURE IN DE-
GREES FAHRENHEIT (F) TO CELSIUS (C) OR
CELSIUS (C) TO FAHRENHEIT (F).
TYPE 1 FOR F TO C OR TYPE 0 FOR C TO F
?0
WHAT IS THE TEMPERATURE IN C?
37
THE TEMPERATURE IN F IS 98.6
END AT LINE 200
```

Go to Frame 10.

---

10   Now let's look at several new statements that will give you far greater computing capability. These statements and those that you used previously will be combined in a variety of ways to show you new programming concepts.

The new statements to be used in this unit are listed in Table 11-4. Review them briefly now.

Table 11-4. BASIC Programming Statements

Statement/ Function Form	Description
FOR.....NEXT	Establishes a program loop with counter. Repeats all the statement between FOR and NEXT as specified.
	Example: For X = 1 to 5
	PRINT X*X
	NEXT X
	Prints the square of the numbers 1 through 5.
READ X	Obtains and assigns the value of X from a DATA statement.
DATA	Specific values to be assigned to variables in a READ statement.
	Example: READ A,B
	DATA 7, 18
	Assigns A = 7, B = 18
INT (A)	Give the integer or whole number value of A only.
	Example: INT(7.6)
	produces the value 7.
ABS (B)	Gives the absolute value of number B. Removes a negative sign.
	Example: ABS(-4.53)
	produces the value 4.53.
RND (1)	Generates a random number of some value between 0 and 1.
	Example: RND(1)
	may produce the number
	.1489305.

Go to Frame 11.

Loops

11   A *loop* is a self-repeating sequence of statements in a program. The loop performs the sequence of instructions more than once before proceeding with the remainder of the program. When the program is first run, the sequence of instructions in the loop is executed once. Then the computer goes back and executes that same sequence again and again as called for by the particular application. Different data is used for each time through the loop. Once the loop has been executed, the program continues.

Loops are used in programming to perform repetitive or iterative operations. For example, a loop will allow you to make the same calculation with a particular formula but with new data each time through the loop. The result will be a tabular output. Loops are also used for counting, sorting, and comparing lists of data.

One way to create a simple loop is to use the GOTO statement you learned earlier. The following program illustrates the basic concept.

(continued next page)

```
10      LET A = 1
20      PRINT A
30      A = A + 1
40      GOTO 20
50      END
```

In this short program, the first statement causes the variable A to be assigned the value of 1. In the next line of the program, the value of A is printed. In line 30, A is given a new value. The expression A = A + 1 is not an algebraic expression and should not be interpreted that way. This is just the notation used in programming to assign a new value to A. Here the new value of A is simply the previous value of A plus 1. This means that after this line is executed, 1 is added to the previous value and the new value of A is given. In this case, 1 will be added to the previous value of 1 to create the new value of 2.

In a computer program, X is given a value of 5. The instruction X = X + 2 is executed. The new value of X is:

    a. 2
    b. 3
    c. 5
    d. 7

---

12   (d. 7)   Referring again to the program in Frame 11, the GOTO statement in line 40 creates the loop. Notice that it tells the computer to go to line 20. By going back to line 20, the loop is formed. The program will again execute the instructions in lines 20, 30, and 40. In this case, the computer will print the new value of A (2). It will then increment the value of A again by 1 and then loop back to print it. The computer will continue to execute this series of instructions.

This program prints the sequential numbers 1, 2, 3, 4, and so on. The main problem with this particular program is that there is no end to it. The computer will continue to execute the statements in lines 20, 30, and 40. The computer has no way to get itself out of this loop. The computer will continue to increment A and print it. The result will be a continuing incremented number being printed on the computer screen.

To overcome this problem, we must introduce an additional statement that allows the program to end it. This can be done by using the decision-making instruction IF THEN previously described in Table 11-1.

```
10      LET A = 1
20      PRINT A
30      A = A + 1
40      IF A >5 THEN GOTO 60
50      GOTO 20
60      END
```

This program is similar to the program in Frame 11 but with the difference being that we have added an IF THEN statement to limit the program loop and allow the computer some way to escape from it. Looking at line 40, the IF A >5 statement tells the computer that once A is incremented enough times and becomes greater than 5, the computer will then go to line 60. As you can see, line 60 is an END instruction that stops the program. If, however, the program status is not such that A is greater than 5, the program will simply loop back to line 20 where the A is printed and again incremented.

Now look at the program. Again A is assigned the initial value of 1 in line 10. The PRINT statement in line 20 causes 1 to be printed. The statement in line 30 increments the value of A. The IF THEN statement in line 40 tests the value of A to see if it is less than or greater than 5. If the value is 5 or less, then the GOTO statement in line 50 simply causes the program to be repeated. At some point in the program, the value of A will exceed 5. At that time, the program will end. As you can see, the value given to the IF THEN statement in line 40 will determine how many numbers are printed.

To print the numbers 1 through 8, the IF THEN statement in line 40 would contain the expression A > ____.

    a. 7
    b. 8
    c. 9
    d. 10

---

13   (*b.* 8)   BASIC provides another statement that makes it even easier to create a loop. This is the FOR NEXT statement. The next program shows how it is used.

```
10      FOR A = 1 TO 5
20      PRINT A
30      NEXT A
40      END
```

The statement in line 10 assigns A the initial value of 1. It also indicates that 5 will be the final value of A. In any case, the first time the program is run, A is made equal to 1. Line 20 then causes A to be printed. The NEXT statement in line 30 causes A to be incremented automatically. Here you don't have to use the A = A + 1 relationship. The NEXT statement increments A automatically and, at that time, tells the program to loop back to line 20. Here the next higher value of A is printed. Again, the NEXT statement increments A and loops to line 20 to print it.

The NEXT statement increments and tests the value of A to see if it is less than 5 or the final value designated by the FOR statement in line 10. If A is not yet at that final value, it causes the program to loop to the previous statement in

(continued next page)

line 20. As soon as the value of 5 has been reached and printed, the program ends. This program prints the numbers 1 through 5 each on a separate line just as the program in Frame 11 did. As you can see, the program using the FOR NEXT statement is shorter because it has two less statements.

To be sure that you understand the FOR NEXT statement, analyze the following program and write the output results. Go to Frame 14.

```
10    FOR X = 1 TO 10
20    PRINT X, X*X
30    NEXT X
40    END
```

14　This program prints the values of X from 1 to 10 and their squares. Remember that the square of a number is that number multiplied by itself. The computer output then is a tabular listing of the value of X and the corresponding square value. By using one PRINT statement and a comma between the quantities to be printed, both the number and its square are printed on the same line.

1	1
2	4
3	9
4	16
5	25
6	36
7	49
8	64
9	81
10	100

Go to Frame 15.

Inputting Data

15　Computers solve problems by performing some type of calculation. The calculation may be an algebraic formula or simply a procedure for rearranging data in a certain way. In either case, the problem requires data. For example, you must supply variables to a formula, or you must supply data that is to be manipulated in some way.

When programming in BASIC there are several means of providing data to the program. You have already learned two of these in the previous unit. The LET statement causes a particular variable to be assigned some value. For example, when we say LET A = 14, we are telling the computer that whenever it encounters the variable A in a program, use the number 14. Some BASICs simply allow you to say A = 14. Such an assignment statement would appear as shown in a BASIC program.

```
10      L = 27
20      J = 35
30      LET L*J = F
40      PRINT F
50      END
```

When the computer executes this program, the answer 945 is printed.

Another method of giving a BASIC program data is to use the INPUT statement. Previously you saw that the INPUT statement could be used to assign a value to a variable. The computer does this by waiting for the user to enter the desired value whenever it encounters an INPUT statement in a program.

For example, consider this simple program:

```
10      INPUT R
20      LET A = 3.14*R^2
30      PRINT A
40      END
```

To execute this program, you type RUN. When this happens, the program executes the statement at line 10. The INPUT statement then causes a question mark to be written on the screen, asking you to supply it with the value of R you desire. This program is computing the area of a circle by the simple formula pi ($\pi = 3.14$) times R squared. When the question mark appears on the screen, you simply type in the desired value for the radius, R. As soon as you type in the desired value, press the Enter key, and the remainder of the program is executed. The area A is then printed on the screen.

Assume that you type in the value 17 when the question mark prompt is given. The computer will print the value:

    a. 17
    b. 53.38
    c. 907.46
    d. 1029.52

16

(*c.* 907.46)   A whole series of input numbers can be entered with a single INPUT statement. For example, consider the INPUT statement:

```
370     INPUT X, Y, Z
```

This INPUT statement on line 370 allows you to enter three separate variables for X, Y, and Z. When the computer executes this statement, it will give the question mark prompt. At that point, you are requested to enter three variables, separate numbers for X, Y, and Z. You enter the data

(continued next page)

by simply typing in the desired numbers for each separated by a comma. Once you have entered the third value, press Enter and the computer will continue to execute the program. Presumably, the X, Y, and Z values that you entered will be part of the calculation.

Consider the following program:

```
10      INPUT M,G
20      PRINT F = M/G
30      END
```

This program will compute your fuel consumption rate (F) in miles per gallon if you tell it the number of miles you traveled (M) and the number of gallons of gasoline (G) used. Assume that M = 324 and G = 12.

When the preceding program is executed, the INPUT statement is executed first and will immediately give the question mark prompt when RUN is typed. At that point, the program is requesting that you enter the value for M. At this time, you type in the number 324. After you press Enter, the computer will again give you a question mark prompt, indicating that you have another value to enter. In this case, it is the number of gallons used. At this point, you enter 12 for the value of G and press Return. Now the computer runs the remaining part of the program from which it computes the fuel mileage by dividing the miles by the number of gallons. The program stops after it prints the number 27 mpg.

There are two additional statements used to give data to the computer program. These are called the READ and DATA statements. The READ statement assigns data values to letter variables. The DATA statement actually supplies the numbers to the READ statement. The READ and DATA statements are always used together in a program. The simple program that follows illustrates how they are used.

```
10      READ A,B
20      DATA 72,18
30      PRINT A*B
40      END
```

The READ statement causes number values supplied by the DATA statement to be assigned to the variables A and B. The DATA statement in line 20 supplies the numbers 72 and 18. The program, therefore, assigns number 72 to the variable A and the number 18 to the variable B. When this program is executed, the computer prints the value 1296.

The READ and DATA statements are an excellent way to supply a volume of data to a program. In many applications, you may wish to run the same program repeatedly with new data. For example, your program may compute the average grade for a number of exams in a course. For each student, there will be a separate set of grades. By simply using different DATA statements for each student, one program can be used to compute the average for each student. Such a program is shown next.

```
10      READ W, X, Y, Z
20      DATA 80, 92, 77, 97
30      PRINT (W + X + Y + Z)/4
40      END
```

When this program is computed, the average grade printed out is:

a. 86.5
b. 87.3
c. 88.2
d. 91.4

---

17 (a. 86.5)   Go to Frame 18.

INT, ABS, and RND Functions

18 In this section, we introduce three new BASIC functions that are widely used in writing programs. These are the integer (INT), absolute (ABS), and the random (RND) functions. The operations they perform are extremely simple, but they are handy in writing some BASIC programs.

The INT function converts any number into a whole number or integer. As you know, any decimal number is made up of two parts, the integer and the fractional or decimal portion. For example, in the number 17.384, the integer or whole number portion is 17 while the decimal fractional portion is the .384.

To use the INT function, you normally combine it with a PRINT or LET statement. For example, to convert the number just given into its integer value, the following statement is used:

PRINT INT (17.384)

When this BASIC statement is executed, the computer will simply print the number 17.

Another way to use the INT function is in conjunction with the LET statement. For example, you could execute the next statement:

LET A = INT (-7.36)

When the computer executes this statement, it assigns the integer portion of the number given to the variable A. In this case, A is simply set equal to –7.

Another commonly used function is the absolute or ABS function. This function converts any number into a positive value. In many calculations, it is the value of the number

(continued next page)

rather than its sign that is important. Many calculations do not permit the use of a negative sign. The absolute function takes a number and removes the negative sign, if it exists.

Like the INT function, the ABS functions can be used with both PRINT and LET statements. For example, consider this statement:

```
PRINT ABS (-45)
```

When this statement is executed, the computer simply prints the number 45. The negative sign is removed. If the number already is a positive number, the ABS function will simply not affect it. When the statement PRINT ABS (85.2) is executed, the computer will print 85.2.

Note that the number was positive in value and, therefore, is unaffected by the ABS statement. Note also that unlike the INT function, the fractional portion of the number was not disturbed.

What is printed by the code below?

```
LET C = ABS (-51.369)
PRINT C
```

---

19 (51.369) When this statement is executed, the variable C is assigned the value of 51.369. Only the minus sign is affected by the ABS function.

One of the most widely used BASIC functions is the function for generating a random number. The RND function causes a number to be chosen at random by the computer. This allows functions such as coin flipping, die rolling and card shuffling, and other random operations to be accomplished.

Different computers and different dialects of the BASIC language cause a random number to be generated in a variety of ways. You should check your BASIC programming documentation before you use it. The example given here is valid for only some computers.

One way that a computer chooses a random number is to display a multidigit value between 0 and 1. For example, the next statement will generate a new random number between 0 and 1 each time it is executed:

```
PRINT RND (1)
```

When executing this statement, you can never tell what number you will obtain. For example, it may be a seven- or eight-digit number between 0 and 1. For example, executing PRINT RND (1) may return the number .4076231.

In most applications, it will be desirable to use values of numbers other than those between 0 and 1. This can be done by simply multiplying the generated random number by a factor. Normally the number can be multiplied by 10, 100, 1000, or any other factor to make it larger in value. The desired range of numbers will determine the factor.

The simple step-by-step program shown next causes a random number between 0 and 99 to be generated:

```
10      LET X = RND (1)
20      PRINT X = 100*X
```

The first statement causes a random decimal number between 0 and 1 to be generated. The next statement multiplies the number by 100 and prints it. For example, if the random number generated is .850367, the next statement will multiply it by 100 and print 85.0367.

Sometimes it is desirable to get rid of the decimal fractions portion of the random number. Of course, this is done by using the INT function. For the previous example, if only a whole number between 0 and 99 is required, the statement PRINT INT (X) could be used. This would simply get rid of the .0367 portion of the number. The computer would then simply print the number 85.

Write a brief program that will generate a random whole number between 0 and 9.

20   (See program below.)

```
10      LET A = RND (1)
20      LET A + 10*A
30      PRINT INT (A)
40      END
```

Statement 10 causes a random number between 0 and 1 to be generated. Statement 20 causes that fractional number to be multiplied by 10. This creates a number between 0 and 9. Finally, statement 30 removes the fractional part of the number and prints the whole number value. Now let's see how some of these functions can be combined in a program to perform some useful application. Refer to the following program:

```
10      LET N = RND (1)
20      LET N = 10*N
30      LET N = INT(N)
40      IF N > 1 THEN 50
50      IF n > 6 THEN 10
60      IF N = 0 THEN 10
70      PRINT N
80      END
```

This program simulates the rolling of a die. It will give some random integer value between 1 and 6 each time it is executed.

The statement in line 10 generates a random number between 0 and 1. The statement in line 20 converts that random number to some value between 0 and 9. The next statement converts the number into an integer value between 0 and 9. Now the IF THEN statements select only

(continued next page)

values 1, 2, 3, 4, 5, or 6. If the number is greater than 6 or 0, the program loops back to 10 and another random number is generated. Finally, the statement in line 70 prints the value 1, 2, 3, 4, 5, or 6.

Go to Frame 21.

Strings

21   We use letters of the alphabet and short two- or three-letter expressions to define variables. These variables have, in turn, been assigned numerical values using the LET statement or READ and DATA statements. You will find that the majority of variables used in BASIC programs are numerical in value. These variables are then used in various algebraic and other expressions to perform mathematical calculations and other operations.

In some applications, you will also find it desirable to work with letters, words, and, in many cases, whole sentences instead of numerical values. This gives you great flexibility when constructing interactive programs. Instead of the user and the computer exchanging numerical values, they can literally talk with one another by inputting words and expressions, as well as numbers. This is done with string variables. A *string* is any series of characters including letters, numbers, spaces, and even special symbols. You have already seen one way in which you can deal with strings. The PRINT statement can be used to cause the computer to write any expression on the screen. The desired string is simply put between quotation marks and used in conjunction with the PRINT statement. When the statement PRINT "THIS IS A STRING" is executed, the computer will write (THIS IS A STRING).

There is also a method of assigning a specific string to a string variable so that it can be readily identified and used in other computer operations. To do this, letters of the alphabet or multiple-letter mnemonics are used to define variables just as they are used to assign numerical values. However, in order to distinguish between numerical and string variables, string variables are followed by a dollar sign. For example, to designate the string NAME, the following expression is used.

```
LET A$ = "NAME"
```

The string variable can then be used in a variety of ways. For example, the statement PRINT A$ could then be executed. This will cause the string NAME to be printed.

Remember that to assign a string variable you must put the string between quotation marks. Also, recall that any characters, numbers, spaces, or symbols can be used in a string definition.

Multiple-string expressions can be combined with a single PRINT statement, as shown next. Executing the statement PRINT A$, B$ will simply cause the string expressions to be printed one after another with a space between them on the same line.

What will the following program print?

```
10      LET A$ = "BLUE"
20      LET B$ = "SKY"
30      PRINT A$, B$
40      END
```

22

(BLUE SKY)   Another way to assign a string to a variable is to use the INPUT statement as you did before with numbers. For example, INPUT X$ will cause the program to print a question mark on the screen and then wait for you to enter a desired string. At this point, you will simply type in the desired word or expression that will then be assigned to the string variable X$. The program can then use that string in any way it needs.

The program below illustrates an example of how the INPUT statement can be used in a simple string program:

```
10      PRINT "WHAT IS YOUR FIRST NAME?"
20      INPUT A$
30      PRINT "WHAT IS YOUR LAST NAME?"
40      INPUT B$
50      PRINT "HELLO", A$, B$
60      END
```

The first statement in this program causes the question "WHAT IS YOUR FIRST NAME?" to be printed. The question mark appears right below it. This is produced by the INPUT statement that is set up to accept your first name. Typing in your first name causes this string to be assigned to the variable A$.

Assume that you type in JOHN. The statement in line 30 then causes the question "WHAT IS YOUR LAST NAME?" to be printed on the screen. The INPUT statement in line 40 prints a question mark asking you to enter your last name. Your last name is then assigned to the string variable B$. Let's say you enter SMITH. Line 50 then causes the expression HELLO JOHN SMITH to be printed. Note that a comma is used between those portions of the string expressions used in the PRINT statement.

The next program will further illustrate the use of strings.

```
10      PRINT "WHAT IS YOUR NAME?"
20      INPUT A$
30      PRINT "OK" A$
40      PRINT "I AM GOING TO WRITE YOUR
        NAME TEN TIMES"
```

(continued next page)

```
50      PRINT "WHEN YOU ARE READY, TYPE ANY
        LETTER"
60      INPUT B$
70      FOR X = 1 TO 10
80      PRINT A$
90      NEXT X
100     END
```

Read through the program and determine what it does before reading the next paragraph. Go to Frame 23.

---

23   This program first asks for your name. The INPUT statement then lets you enter it and assigns it as a string variable A$. Next, the computer tells you that it will print your name ten times. The input statement on line 60 causes the computer to pause so that you can read the statement printed by line 50. You can then type in any letter that is assigned to B$. This assignment has no meaning. The only value of the INPUT B$ statement is to pause.

When you enter any letter and press the return or enter key, the computer prints your name ten times. This is done by the FOR NEXT loop made up of lines 70, 80, and 90.

As you can see, BASIC is a versatile, easy-to-learn language. You can program almost any application with it. And over the years BASIC has continually improved. The most popular version of BASIC today is Microsoft's Visual BASIC, which combines the BASIC language with development programs that let you program your applications for the Windows GUI. Over 85 percent of all personal computers use the Windows operating system. Visual BASIC allows you to create applications programs with windows, buttons, pull-down menus, and the like without programming. Your program in BASIC can then interface with the useful and familiar Windows interface with a minimum of programming effort.

Go to Frame 24.

BASIC in Embedded Control

---

24   Further evidence of BASIC's versatility is its widespread use in industrial control and embedded microcontrollers. Although BASIC's primary use is in business and engineering applications, it has been widely adopted as a simplified way to perform the I/O operations for monitoring and controlling external devices. Personal computers and their hardened industrial derivatives running BASIC are very popular in plant monitoring and industrial control operations. In addition, the availability of special control versions of BASIC in microcontrollers has made it a popular language for building embedded applications.

The key to BASIC's use in such applications was the inclusion of special statements and commands that allow BASIC to access the computer's memory and I/O ports directly. Control and embedded applications use interfaces to connect the computer to inputs to be read and outputs to be used in the control of external devices. When input/output capability was added to BASIC operations, it became a control language.

The first of such commands to be added were PEEK and POKE. The PEEK statement takes the form PEEK(M) where M is the decimal address of any word in the computer's memory space. In a microcomputer with a 16-bit memory address, you can address $2^{16} = 65,536$ memory locations. Some LOCATIONS may be RAM, others ROM, some memory-mapped I/O addresses, and still others unused. Here are some examples of PEEK.

PRINT PEEK (2047)

This writes the decimal value of the binary number in memory location 2047 on the screen.

LET X = PEEK (32768)

This assigns the decimal value of the *contents* of location 32768 to X.

Just remember that the value of a memory word depends on its bit length. With an 8-bit memory word, the decimal value may be any value between 0 and 255.

PEEK lets you look at memory contents or perform input operations if memory-mapped I/O is used. PEEK also lets you interface a BASIC program with a machine or assembly language program.

The PEEK command uses data and addresses in:

    a. hex
    b. decimal
    c. binary
    d. any number system you choose

---

25   (*b.* decimal)   The POKE command causes a desired value to be stored in memory. It has the form POKE X,Y where X is a decimal memory address and Y is a decimal value to be placed into X. For example, POKE 127, 100 stores the decimal value 100 at decimal memory address 127.

POKE lets you store values in RAM under control of a BASIC program. It also allows you to perform output operations if memory-mapped I/O is used. Be sure to know the memory and word sizes so that valid values can be used.

What value is returned by the following program?

(continued next page)

```
10    POKE 4095, 80
20    LET A = PEEK (4095)
30    LET X = A + 15
40    PRINT X
50    END
```

25 (95)   Two popular examples of microcontrollers that use BASIC are Intel's widely used MCS-51 and Microchip Technology's PIC series. (See Unit 7.) A version of the Intel MCS-51 microcontroller known as the 8052 contains BASIC-52, a BASIC interpreter on-chip in the program ROM. A BASIC program is created and stored in data RAM or PROM. The interpreter executes the program.

The Microchip Technology PIC16C56 microcontroller is used by the company Parallax Inc. to create the BASIC Stamp. This tiny microcontroller contains a control version of BASIC, known as PBASIC, in program ROM. An external EPROM used for program memory stores the BASIC program. PBASIC, an interpreter, executes the control program. Both versions of BASIC contain special statements and commands that greatly simplify and facilitate I/O operations that are the heart of all embedded controllers. Some examples are given next. The MCS-51 and PIC series micros have a Harvard architecture with separate data and program memories. These are internal to the chip and, in some cases, external.

Go to Frame 26.

BASIC-52

26   Intel's BASIC-52 contains virtually all of the statements and commands covered previously in this unit or some variation. In addition, it implements many additional commands that perform useful I/O operations or more advanced processing functions. Some examples follow:

.AND	Performs the logical AND function.
CALL	Branches to an assembly language program also in program memory.
CBY	Retrieves a word from program memory, similar to PEEK.
DBY	Retrieves or assigns a value from internal data memory.
DO WHILE	Executes all statements between DO and WHILE, if any, until some given condition is *not* met.
GET	Assigns the ASCII value from an external keyboard to some variable.
GOSUB	Jumps to a subroutine at a given line number.
LOG	Computes the natural logarithm of a number.
ONEX1	Recognizes an interrupt on pin 13 of the 8052 and jumps to a subroutine at a given line number.
PH1	Similar to PRINT but displays values in four hex digits. Example: A27Dh; the lower case "h" indicates a hex value.
PWM	Outputs a pulse width modulated signal to an output port. The format is PWM X, Y, Z, where X, Y, and Z are numbers designated a time period for the high-pulse duration (X), low-pulse duration (Y) and number of output cycles. The decimal values for X, Y, and Z are given in 8052 clock cycles equal to 12/clock crystal frequency.
RETI	This statement must be used at the end of a subroutine that was executed as the result of an interrupt. It returns the program to the line after the ONEX1 statement.
SIN	Gives the trigonometric sine value for an angle in degrees.
SQR	Computes the square root of a number.
TIMERN	N is 0, 1, or 2. This command retrieves or assigns a value to the 8052's timer registers.
XBY	Retrieves or assigns a value in external data memory. Used for both input and output to external devices.

Here is a simple program that accepts a pushbutton input on the interrupt pin 13 and in response outputs a binary 1 bit to I/O port E200h to turn on an external LED. The program loops until a binary 0 at the switch input occurs and then turns on the LED.

```
10      ONEX1 50
20      DO
30      WHILE 1 = 1
40      END
50      XBY(E200h) = 01
60      RETI
```

The ONEX1 statement enables external interrupt 1 on the 8052 microcontroller. It will cause the processor to finish any statement in progress to be completed and then it will branch to a subroutine at line 50 if the interrupt occurs. The DO WHILE statements wait for the interrupt in an endless loop. When the input switch is pressed, the interrupt line goes low. The DO WHILE detects this and the program branches to the subroutine at line 50. The subroutine is just an XBY output statement that sends the hex value 01 to output port with the address E200h. The hex value 01 is 00000001 in binary. The binary 1 line is connected to an external LED that turns on. The RETI returns from the subroutine to the main program.

Go to Frame 27.

BASIC Stamp

27 Although Intel's BASIC-52 is quite traditional, Parallax's PBASIC is not. It works like BASIC and uses some of the more common BASIC statements and commands, but it has been simplified and enhanced to make it a true control BASIC for embedded applications. It is more limited in computing ability than BASIC-52 because it performs integer (whole numbers) math only. Decimal points and fractions are ignored. For example, LET X = A/B, where A = 7 and B = 2 gives the value 3, not 3.5, as would BASIC-52. In most cases, this is not a disadvantage in many embedded applications.

Listed below are some of the special commands of PBASIC.

BUTTON	Defines one of the BASIC Stamps eight input pins to receive a binary input from a pushbutton. It also allows you to debounce the pushbutton and indicate whether the closed or depressed state of the button in binary 0 or 1.
HIGH	Makes a designated output pin high (+5 volts).
INPUT	Assigns an I/O pin to be an input.
LET	LET in PBASIC is used to assign values and do basic math. But it also allows you to do logical operations such as AND, OR, NAND, NOR, XOR, and XNOR.
LOW	Makes a specific pin low or binary 0 (0 volts).
OUTPUT	Assigns one of the eight I/O pins to be an output.
PAUSE	Waits for a period of time given in milliseconds from 0 to 65535.
POT	Reads the value or setting of a 5K to 50K potentiometer connected to a specific I/O pin. Good for reading resistive sensors like thermistors and photocells.
PULSEIN	Measures the duration of an input pulse on a stated pin in 10 µS increments.
PULSEOUT	Generates a pulse on a designated pin in 10 µS increments.
PWM	Same as the PWM command discussed earlier.
READ	The PBASIC READ command is unlike the standard READ command discussed earlier. Instead it is like PEEK or BASIC-52's CBY or XBY commands.
SERIN	Sets up an I/O pin to read a serial data train at a specific baud (data) rate.
SEROUT	Sets up an I/O pin to be an output and to send a serial data stream at a given baud rate.
SOUND	Sends a square wave to a designated I/O pin to create a tone in an external speaker. Frequency from 94.8 Hz to 10,550 Hz can be created.

The following is an example program that flashes an external LED connected to the output port pin 7. The BASIC Stamp uses a PIC16C56 microcontroller as described in Unit 7. It has a single 8-bit I/O port that may be set up for either input or output as determined by the BASIC statement executed.

```
Flash:
   Output 7
   High 7
   Pause 500
   Low 7
   Pause 500
   Goto Flash
```

Notice that no line numbers are used and that you may also use lowercase letters for the statements. The first line establishes "Flash" as a subroutine as denoted by the colon. Output 7 makes pin 7 an output pin. Then High 7 makes the output high or +5 volts. This turns on an external LED connected to pin 7. The program then pauses 500 milliseconds or one-half second. The next statement, Low 7, makes pin 7 go low or to ground, turning off the LED. The next, Pause 500, waits another 500 milliseconds. Finally, the Goto Flash causes the program to loop back to the beginning and repeat. The LED continues to flash off and on every one-half second.

Although the BASIC Stamp is by far the most popular embedded BASIC microcontroller, several other manufacturers offer similar units, all optimized for dedicated monitoring and control operations.

Answer the Self-Test Review Questions on the diskette before going on to the next unit.

Advanced Microprocessors

1   As you learned in previous units, microprocessors use a fixed word size. The first successful commercial microprocessors were 8-bit units; all of the first personal computers were based upon them. Today 8-bit microprocessors are still the most popular and widely used size for embedded microcontrollers.

Sixteen-bit microprocessors came along in the late 1970s but were not widely adopted until the 1980s when they became the heart of the second generation of personal computers. Over the years, 16-bit microprocessors and microcontrollers have also been developed for embedded ap-

(continued next page)

plications. Today, the most popular microprocessors for personal computers have 32 bits. Further, 64-bit microprocessors are now widely available and are used primarily in advanced engineering workstations and network servers. Sixty-four-bit personal computers will be the norm in just a few years.

In this unit, you will learn about the benefits of larger word length microprocessors and discover the techniques that make them more powerful. You will review the most popular 16-bit microprocessors, which were the basis for most of the newer 32-bit and 64-bit designs. The Intel X86 series of microprocessors will discussed as they relate to personal computers.

Table 12-1 traces the development of the microprocessor over the years. Scan this table before you proceed. Note that the term *MPU (microprocessing unit)* is sometimes used in place of the word *microprocessor* or *central processing unit (CPU)* in this table and the text that follows.

Table 12-1. Historical Perspective of Microprocessors

1971	First "computer-on-a-chip" developed by Intel. Known as the 4004, a 4-bit microprocessor, it was used in early electronic calculators.
1973	Intel develops the 8080 microprocessor, the first and most widely used microprocessor.
1975	Motorola, Signetics, MOS Technology, Texas Instruments, National Semiconductors, and Zilog develop 8-bit microprocessors.
1975	First personal computers announced. Computers based upon the Intel 8080, MOS Technology 6502, Motorola 6800, and Zilog Z80 are the most popular.
1977	Texas Instruments and National Semiconductor announce 16-bit microprocessors, the 9900 and PACE respectively, neither of which became popular.
1978–79	Intel introduces the 8086, and Motorola announces the 68000 16-bit microprocessors, both of which become enormously popular.
1981	The IBM PC is introduced using a version of Intel's 16-bit 8086 (the 8088).
1984	Apple introduces the Macintosh using the Motorola 68000 16-bit microcomputer.
1984	IBM introduces the PC AT using Intel's advanced 16-bit micro called the 80286 or 286.
1989	Intel introduces the first popular 32-bit microprocessor, the 80386 or 386. It is widely adopted by PC manufacturers. Motorola introduces its 32-bit MPU, the 68030, which is used in Apple's more powerful Macintosh PCs.
1991	Intel launches the most powerful 80486 or 486 32-bit MPU. Motorola announces a competitive 32-bit MPU called the 68040. Both are widely used in IBM PC-compatibles and Apple Macintosh.
1993	Intel introduces the Pentium 32-bit MPU.
1994	Motorola, with IBM and Apple, introduces the 32-bit Power PC. It is used by Apple in the new Macintoshes and in some IBM workstations.
1996	Intel introduces the Pentium Pro, a more advanced Pentium model. Motorola introduces faster, more powerful Power PC models. Semiconductor companies AMD and Cyrix introduce Pentium-compatible processors called the K6 and 6X86 respectively.
1997	Intel introduces the Pentium II, a still more powerful Pentium with a different bus structure.
2000	Estimated introduction of Intel's IA-64 (code named Merced) advanced 64-bit microprocessor.

The pace of microprocessor development, introduction, and adoption into new products has been absolutely furious over the past 25 years and, in fact, the pace has quickened in the past decade. We owe this phenomena to the basic high volume and competitive nature of the semiconductor and personal computing businesses. But just as important is the semiconductor manufacturing technology that permits larger, faster, more complex chips to be made at lower and lower prices. The techniques of making a microprocessor, memory, or complex communications chip have improved greatly over the years. Early integrated circuits introduced in the early 1960s had only a few transistors. As manufacturing methods improved, it became possible to make transistors and other components smaller and smaller, thus permitting circuits to be larger and more complex in the same or less space on the silicon chip. Today's most advanced microprocessors such as the Pentium contain millions of transistors—and memory chips contain tens of millions of transistors. Modern semiconductor manufacturing techniques permit transistors with a size of 0.25 to 0.35 μ (micron) to be made. (A micron is one-millionth of a meter. A typical human hair diameter is in the 100 to 400 μ range.) This trend toward even smaller transistors will continue, allowing larger circuits to be created on smaller chips of silicon. Smaller transistors are also faster, so that they process data at ever higher rates of speed.

In the early days of integrated circuits, one of Intel's founders, Gordon Moore, predicted that the density of transistors and circuits on ICs would double every 24 months. This was later revised to every 18 months. Semiconductor technology has kept pace with this prediction. Known as Moore's law, this prediction is expected to be true for years to come, when it will be possible to put over one billion transistors on a chip by 2010.

Which of the following is *not* true?

    a. Most microcontrollers still use 8 bits.
    b. Smaller transistors are slower.
    c. Most personal computers use a 32-bit microprocessor.
    d. A transistor can be made smaller than one micron.

---

2   (*b.* Smaller transistors are slower.)   Go to Frame 3.

Features That Make
Microprocessors More Powerful

3   Although most microcontrollers still use an 8-bit architecture, virtually all personal computers use 32-bit microprocessors. Microcontrollers are typically used in relatively simple monitor and control operations. They do not process large amounts of data or use data that involves very large or very small numbers. Furthermore, processing operations are usually simple math and logic, plus a great deal of I/O.

Personal computers, on the other hand, are general-purpose machines that must be flexible enough to handle huge quantities of data quickly. They must use fancy color and graphics, handle multiple tasks concurrently, and deal with multiple I/O peripheral devices such as disk drives, printers, mouse, video monitors, modems, and network interfaces. PC's must be fast and flexible. They require far more computing power than a microcontroller.

But what is "computing power"? What makes one processor or computer better than another? The answer is:

- Greater speed
- Wide data-value ranges
- Larger memory sizes
- Computing flexibility

An important factor in achieving all of the above is word size, although other factors come into play as well. Let's discuss each of these factors in more detail.

Speed is achieved by using faster logic circuits (lower propagation delays) and running all circuits at a higher clock frequency. In integrated circuits, faster logic circuits are built by using smaller transistors packed closer together on the chip. With current microprocessors using 0.25 to 0.35 μ transistors, clock speeds of 200 to 450 MHz and higher are readily achieved. Speed is also a function of the number of layers or levels of logic. The simpler the circuits and the fewer the levels, the less the overall propagation delay and the faster the computation. In coming years as transistors become even smaller and as ways are found to shorten distances between them, common clock speeds up to 1 GHz will be obtained.

Which of the following factor does *not* affect CPU speed?

    a. transistor size
    b. transistor spacing
    c. levels of logic
    d. number of transistors

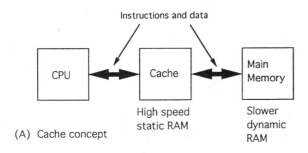

(A) Cache concept

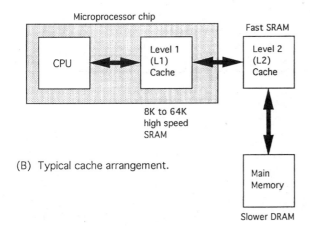

(B) Typical cache arrangement.

Fig. 12-1. Cache memory. (A) Cache concept. (B) Typical cache arrangement.

4 (*d.* number of transistors)   Faster processors are important to achieve super high-speed math, logic, and other operations. But just as important is memory speed. Processors access instructions from memory sequentially, so the memory must be able to keep up with the fast ALUs in the microprocessors. If they cannot, processing slows to the memory speed (access time). In general, memories have not kept up. Memories have gotten bigger and cheaper but not necessarily faster. However, tricks like cache memory have been developed to overcome this problem.

Cache memory is a small, fast, semiconductor memory positioned between the microprocessor and the main RAM, which is usually slower but cheaper dynamic memory (DRAM). Refer to Figure 12-1A. DRAM has a typical access time of 60 to 70 nanoseconds. The MPU can access RAM much faster, usually in only a few nanoseconds depending upon the clock speed. Since the DRAM cannot keep up with the processor, the processor must wait, considerably slowing computing time.

The cache memory is usually made with fast static RAM (SRAM). Its size varies, but is usually in the 4K to 512K word range. It has an access time that more closely matches the MPU. The system is set up so that a portion of the program and data currently being used is brought in from DRAM and temporarily stored in the cache SRAM. In this way, the MPU accesses instructions and data sequentially from the faster SRAM and overall throughput increases dramatically. The MPU thinks that it is getting all of its instructions and data from the fast RAM.

If the MPU attempts to access an instruction or data word not in cache, it does get it from the slower DRAM and computing continues, but at a slower pace. The system then reacts by updating the segment of the program in cache. Optimized cache systems manage to keep that portion of the program currently in use in the cache.

Cache memory has been used for years in mainframe computers and minicomputers to increase computing speeds. They were first incorporated into personal computers in the late 1980s, and today they are a standard feature on all 32-bit personal computers. In fact, they are considered to be a primary feature of the microprocessor. Most advanced MPUs have a small (8K–64K) cache RAM on the same chip.

Over the years cache memory systems have evolved into a two-level system like that shown in Figure 12-1B. The small, fast cache on-chip with the MPU is known as a level 1 (L1) cache. The larger cache is called the level 2 (L2) cache. L2 cache is 128K to 512K words and is made with fast SRAM. Although not as fast as L1 cache, L2 cache runs with 10 to 15 nS access times. L2 cache gets its program segment and data from the main RAM. L1 cache gets

(continued next page)

its instructions and data from L2 cache. The overall result is a 95 percent or greater hit rate. Hit rate is the probability that the instruction or data needed next is in the cache. If it is not, a miss occurs and the MPU goes to the main DRAM to get it. With such a high hit rate, the performance is dramatically improved over the same fast CPU with no cache. In the newer MPUs, separate instruction and data caches improve performance even more.

Bus speed is another factor. The address, data, and control bus lines, along with any connectors, add inductance, capacitance, and resistance. This forms low-pass filters and transmission line effects that slow the movement of data and switching speeds of control signals. Typical bus speeds are 33 MHz and 66 MHz in modern PCs. In some newer PCs, the bus speed is 100 MHz. So, although a microprocessor may be able to achieve a 266 MHz processing speed internally, rarely is data actually moved or processed this fast because the memories and buses slow it down. Over the years, bus speeds have improved considerably, but they are still a key speed limiting factor.

The cache memory overcomes the low speed

    a. DRAM
    b. SRAM
    c. CPU
    d. bus

---

5   (*a.* DRAM)   Word size also affects speed. Processors with 8- and 16-bit registers can deal with larger values by dividing and storing larger numbers into words that the processor can handle. A 32-bit value can be stored as four bytes, and then processed in four-byte increments. But that takes extra time. It is faster to process all bits simultaneously if the processor word size permits.

If data is represented in bytes such as ASCII, larger word sizes (buses, registers) permit you to store and move data faster. With a 32-bit CPU and data bus you can move four bytes simultaneously. If an 8-bit bus can only transfer data at 33 MHz, its speed is 33 megabytes/second or 33 Mbps. With a 32-bit bus, four bytes are transferred at a time, giving a data rage of 4×33 MB or 132 Mbps. We normally say that microprocessors with larger word sizes and buses have greater *throughput*. Throughput refers to the total amount of data processed or communicated over a specific period of time.

Another advantage of larger word lengths is that very large or very small numbers can be more easily represented. The maximum and minimum values you can represent with different word sizes are:

Word length	Largest	Smallest
8 Bits	255	.0039
16 bits	65,535	.000015259
32 bits	4,294,967,296	2.328306×10^{-10}
64 bits	1.844674×10^{19}	5.421011×10^{-20}

You can easily see the value of using more bits to represent larger and smaller numbers. Even larger and smaller values can be achieved by using floating-point numbers.

Larger word sizes also permit larger memories to be addressed. Most 8-bit processors have a 16-bit address word, allowing $2^{16} = 65,536$ words to be accessed. Sixteen- and 32-bit processors usually have 24- or 32-bit address words, letting you use up to $2^{24} = 16,777,216$ (16 meg or 16M) words or $2^{32} = 4,294,967,296$ (4 giga or 4G) words of instructions or data.

Most of the newer microprocessors also support *virtual memory*. Virtual memory is the process by which the CPU can address not only huge volumes of main memory such as RAM and ROM, but also can address programs and data stored on a hard disk. The operating system makes the CPU think that all of the RAM, ROM, and disk drive memory capability appears as one continuous and usable memory space. Virtual memory systems permit extremely large programs and databases to be accommodated.

A list of 32-bit numbers could be added by an 8-bit MPU.

a. True
b. False

6 (*a.* True)   Another factor in computing power is computing flexibility. This refers to the number and type of instructions used in the processor. One of the more subtle reasons behind the greater power of a 32- or 64-bit microprocessor is the instruction sets. These microprocessors not only have more instructions but also have more powerful and sophisticated instructions. A good example of this is the instructions used for multiplication and division commonly available in every 16-, 32-, or 64-bit microprocessor. Although 8-bit CPUs can perform addition and subtraction, many cannot perform multiplication and division directly. Instead, these operations have to be programmed. Usually subroutines are written to perform multiplication and division. The same is true for many other math operations. Floating-point operations can be programmed on microprocessors without floating-point hardware, but they are much slower. Advanced CPUs have built-in floating-point hardware supported by floating-point instructions.

In general, any process can be programmed on any size microprocessor. But the resulting multiple instruction routines take up valuable memory space and require a large amount of time to execute. In a 32-bit microprocessor, many advanced operations can be performed by simply writing and executing a single instruction. There are many sophisticated instructions available in 32-and 64-bit microprocessors. These include special instructions for proc-

(continued next page)

essing BCD numbers or single instructions for causing blocks of data to be moved from one place to another. In general, these instructions perform operations with additional hardware that would ordinarily have to be programmed with multiple instruction subroutines in 8-bit microcomputers.

Finally, many 32- and 64-bit microprocessors have more and powerful addressing modes. Addressing modes, as you recall, refer to the ways in which an instruction or a piece of data in memory can be located.

As microprocessors have developed over the years, they have become more powerful because word sizes have increased, clock frequencies have increased, larger memories can be addressed, and instruction sets have gotten larger. The architecture of microprocessors has become far larger and more complex to support all of these improved features. Microprocessors such as this are referred to as *complex instruction set computers (CISC)*. As the number and complexity of instructions have increased, logic speed increases have been necessary just to maintain high levels of throughput. The Intel 486, the Pentium, and the Motorola 68040 microprocessors are good examples of CISC.

Another school of thought about the architecture of computers has also developed over the years. It is the antithesis of CISC. Known as *reduced instruction set computers (RISC,* pronounced "risk"), this architecture seeks to reduce the number of instructions, simplify the circuitry, and, in turn, greatly increase computing speeds. Many of the newer more powerful microprocessors, especially those used in engineering workstations and network servers, use a RISC design.

RISC microprocessors are characterized by:

- Very small instruction sets (usually <100 or even <50).
- A large set of general-purpose computing registers
- Simple circuitry
- Very high speeds
- Large memory address capability

Some examples of RISC designs are the Motorola 601, 603, 604, and G3 Power PC microprocessors; Digital Equipment Corporation's (now part of Compaq) Alpha; and Sun Microsystems' Ultra SPARC. All are 64-bit designs with clock speeds up to 600 MHz. RISC designs have also been incorporated into smaller microcontrollers. Microchip Technology's PIC16C5X, which you learned about in a previous unit, is an example of an 8-bit RISC design.

Although RISC microprocessors require many operations to be programmed, unlike similar operations in CISC designs, the usually higher processing speeds typically offset the longer subroutines required. Further, RISC designs have become more complex over the years, with most of the new 64-bit CPU incorporating floating-point instructions.

Which is usually faster?

a. CISC
b. RISC

7 (b. RISC) Another method of increasing the performance of a microprocessor is to use what is called *pipelining*. Pipelining is the process of overlapping the fetch, decoding, and execution of instructions. In other words, the fetch, decode, and execute operations are done in parallel with one another. Figure 12-2A shows the standard sequential process in most CPUs. Pipelining is illustrated in Figure 12-2B. Separate circuits are used to pre-fetch and pre-decode a stream of instructions so they can be executed sequentially much faster. Other operations that take extra time, such as address formation, can also be pipelined. Today, virtually all advanced microprocessors use some form of pipelining to speed up processing.

Go to Frame 8.

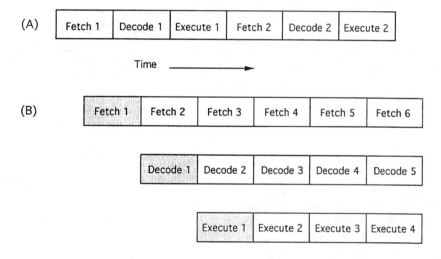

Fig. 12-2. (A) The fetch-execute cycle of the typical traditional CPU vs. (B) the overlapping fetch-execute cycles of a pipeline CPU.

The 8086/8088 Microprocessor

8 One of the most popular and widely used 16-bit microprocessors is the Intel 8086. A modified version of the 8086, known as the 8088, was selected for use in the original IBM PC and its many clones. In this section, we will discuss the architecture and the operation of the 8086/8088 series of microprocessors. Although no longer used in modern PCs, the architecture of this series was the

(continued next page)

basis for later designs like the 286, 386, 486, and Pentium microprocessors. A version known as the 80186 is still used in some embedded applications.

The primary difference between the 8086 and the 8088 is the size of their data buses. The 8086 can transfer 16 bits at a time between memory and I/O circuits, but the 8088 has an 8-bit bus. The 8-bit data bus allows only byte-wide accesses to memory and I/O. Otherwise, the two CPUs are virtually identical. For example, software written for one will run on the other.

The 8086/8088 microprocessors are like most other microprocessors in that they are complete central-processing units. As you may recall, a CPU consists of the general-purpose registers, the arithmetic logic unit (ALU), and the control unit. The control unit fetches and interprets the instructions while the ALU and general-purpose registers carry out the data transfers, arithmetic, logical, and other operations specified by the op codes. In the 8086/8088 microprocessors, all of the CPU circuitry is generally referred to as the *execution unit (EU)*.

In addition to the execution unit, the 8086/8088 microprocessors also contain a new section designated the *bus interface unit (BIU)*. The bus interface unit controls the flow of instructions from the internal systems bus and the execution unit, as shown in Figure 12-3. The main purpose of the BIU is to fetch a number of instructions and temporarily store them, and to make available a stream or pipeline of ready-to-execute instructions for the EU.

In most CPUs, the basic operation that takes place is fetching and executing of instructions. The control unit fetches an instruction from memory, stores it, interprets it, and then directs the ALU to carry it out. This alternate fetch-execute is repeated again and again as the instructions in a program are executed. The important point about this cycle is that the CPU does not typically fetch another instruction until the execution of the previous instruction is complete. As a result of this, the ALU and general-purpose registers waste much of their time waiting for instructions to be fetched and decoded. The 8086/8088 microprocessors eliminate this wasted time. Because of the separate BIU and EU sections, fetch and execute operations can be performed simultaneously. This parallel fetching and executing gives the 8086/8088 extremely high performance over other CPUs. The overall result is higher speed or greater throughput. This unique feature makes the 8086/8088 microprocessors extremely powerful.

The 8086 uses pipelining.

    a. True
    b. False

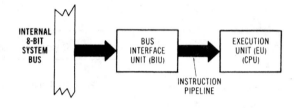

Fig. 12-3. **Parallel pipeline architecture of the 8086/8088 CPU.**

9   (*a.* True)   The primary architectural feature of a microprocessor that describes how the CPU performs is determined primarily by its register organization. The

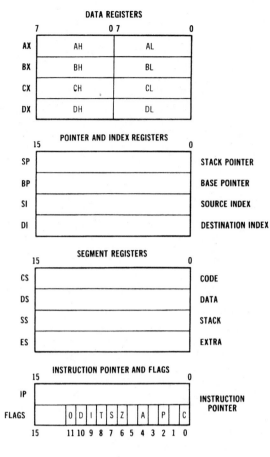

DATA REGISTERS

POINTER AND INDEX REGISTERS

SP — STACK POINTER
BP — BASE POINTER
SI — SOURCE INDEX
DI — DESTINATION INDEX

SEGMENT REGISTERS

CS — CODE
DS — DATA
SS — STACK
ES — EXTRA

INSTRUCTION POINTER AND FLAGS

IP — INSTRUCTION POINTER

Fig. 12-4. The 8086/8088 register structure.

8086/8088 microprocessors have eight general-purpose registers as shown in Figure 12-4. Each register is capable of storing a 16-bit word. All of these registers fit the general definition of an accumulator register as described earlier. Any one of these eight registers can, in effect, be used as an accumulator.

The general-purpose registers are divided by function. Even though each register is capable of being used as an accumulator, typically the registers are reserved for specific operations. The first four registers are usually designated data registers, while the second four registers are used for pointer and index operations. The data registers are divided into 8-bit groups so that each of the four registers can be used to contain a 16-bit word or two 8-bit words. In other words, the high and low halves of each of the four data registers can be used separately in some arithmetic-logic or data transfer operation. For example, in Figure 12-4, the BX register has an 8-bit lower half (BL) and an 8-bit higher half (BH).

The pointer and index register portions of the general-purpose registers are used primarily in addressing operations. The SP and BP registers are the stack-pointer and base-pointer registers, respectively. Most microprocessors have only a single stack-pointer register. The 8086/8088 microprocessors have the stack pointer plus an additional base-pointer register that is also used for stack references. Recall that a stack is a linear sequence of memory locations used for storing subroutine parameters, subroutine return addresses, and other data temporarily saved during the execution of a subroutine or interrupt. The stack pointer is used like other stack pointers in that it points to subroutine and interrupt addresses. The base pointer register, however, is available to the programmer for related stack operations or for use in whatever way desired.

The other two 16-bit registers in the pointer and index group are index registers. These are designated the source index (SI) and the destination index (DI) registers. Index registers, as you recall, are used in modifying the address of a data reference instruction. Usually the content of an index register is added to the content of the program counter to form the address where the data is stored. The SI and DI registers can be used for standard indexing operations and both have auto-incrementing and auto-decrementing capabilities. These registers are also used by the unique string manipulation instructions available in the 8086/8088.

Also contained in the 8086/8088 are the segment registers designated CS, DS, SS, and ES. Again, refer to Figure 12-4. These registers are used to form the memory address, which is used to locate any instruction or data in memory. All accesses to memory are made through one of these registers.

(continued next page)

Memory in the 8086/8088 is divided into segments. The segmentation is based on how the CPU uses the memory and the types of information stored there. There are three basic types of information stored in memory. These are programs made up of instructions, data, and the stack. Typically it is desirable in programming to separate or segment the instruction code portion of the program from the data and from the stack area. This is easily done with the segment register concept of addressing.

The four segment registers are used to address each of these different types of information. The CS register is used for addressing the instruction code segment of memory. The DS register is reserved for addressing that segment of the memory holding data. The SS register is used for accessing information in that area of the memory designated as the stack. Finally, an extra segment register is provided to expand on any one of the four. Typically the ES register is used for referencing data.

There are two additional registers in the CPU that are generally known as control registers. See Figure 12-4. One of them is the instruction pointer (IP) and the other is the flag register. The instruction pointer is a 16-bit register that is similar to the program counter in other microprocessors. The unique feature of the instruction pointer in the 8086/8088 is that it points to the next instruction to be fetched. Normally the program counter in a CPU points to the next instruction to be executed. Because of the use of the bus interface unit that "looks ahead" by fetching a number of instructions prior to execution, the instruction pointer keeps track of the next instruction to be fetched.

The IP register points to the next instruction to be:

    a. fetched
    b. executed

---

10   (*a.* fetched)   Finally, the 8086/8088 has a flag register. There are nine flags used to indicate various status and control functions in the CPU. They are tested by various instructions and are used to cause changes in program control.

The 8086/8088 CPU can address one megabyte of memory. With a 20-bit address word, the CPU can access a total of $2^{20} = 1,048,576$ bytes. This, of course, may be RAM or ROM. As indicated earlier, the memory is also divided into segments. These segments are reached by forming an address with the segment registers. There are assumed to be sixteen 64K segments. The segment registers can address all 64K bytes in any one segment.

Since none of the registers in the 8086/8088 are twenty bits in length, you may wonder how the 20-bit memory address is created. This is done as shown in Figure 12-5. The basis for the 20-bit memory address is the segment register content to which four additional bits are appended. The resulting 20-bit word is applied to an adder. The other input

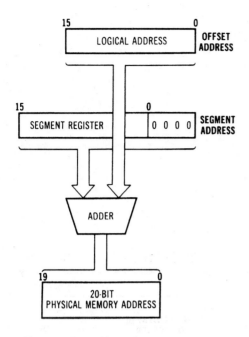

Fig. 12-5. Formation of the 20-bit address.

to the adder is a 16-bit word called the offset address. This offset address can come from a variety of sources, including the instruction pointer, the pointer and index registers, plus a displacement that may be part of the instruction. When the segment address is added to the offset address, the 20-bit memory address is formed. This is the 20-bit word that appears on the address bus that is used to address either RAM or ROM.

As an example, consider the access of instructions. Here the offset address comes from the instruction pointer. The instruction pointer content is added to the segment address to form the 20-bit memory address. Recall, also, that certain areas of memory are designated for instructions (codes), data, or stack. In most cases, then, the instruction pointer contents will be added to the code segment register (CS) to form the memory address to address an instruction.

To form a 20-bit address to access an instruction, the segment address made up from the CS segment register is added to the content of the:

a. register AX
b. base pointer
c. instruction pointer
d. ES register

11 (*c.* instruction pointer)   To address data words, the segment address is added to the offset address which, in turn, can come from a variety of sources or combinations of sources. The offset address may come from the base pointer (BP) register, the SI or DI index registers, or a displacement, which is an address word that is part of the instruction. Different combinations of these are also used to form the offset address. The offset address can be the sum of the base register and an index register; or the sum of the base register and a displacement; or the sum of the base register, an index register, and a displacement. Circuitry in the CPU adds these various address sources to form the offset address, which is then added to the segment address.

As you can see, the addressing capability of the 8086/8088 is extensive. All kinds of address modification schemes can be performed. This provides the programmer with a variety of choices for accessing instructions or data. Importantly, since the 20-bit address word is made up of the contents of various other registers, naturally these registers must be previously loaded with the desired segment, base value, index value, or other value prior to fetching or executing an instruction. A variety of instructions are used in the 8086/8088 to prepare these registers with the proper content.

The number of types of addressing modes used in the 8086/8088 are summarized in Figure 12-6. These include

(continued next page)

register, immediate, indirect register, direct addressing, indexed, relative, base, and base-indexed addressing. Remember that when the indexed, relative, base, or base-indexed forms of addressing are used, a computation takes place to add together either an index register, a base register, or an instruction displacement address to form the offset. This, in turn, is added to the segment address to form the final 20-bit address.

Now let's look at the instructions for the 8086/8088. The instruction format is summarized in Figure 12-7. An instruction can contain anywhere from 1 to 6 bytes, or 8 to 48 bits. All instruction op codes occupy one byte. Besides

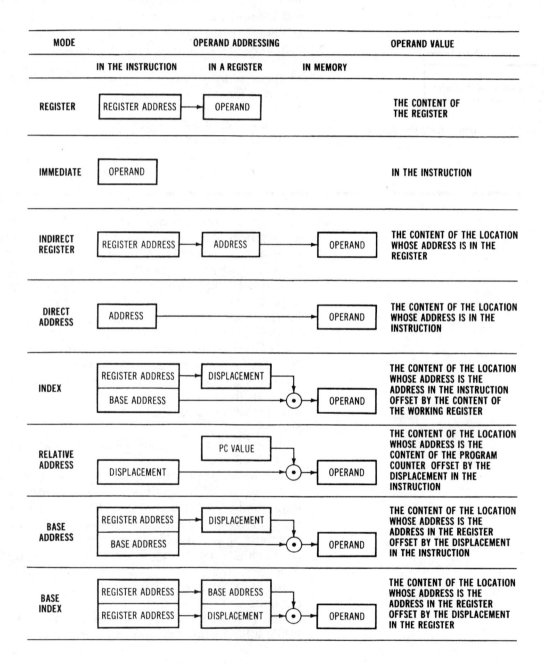

Fig. 12-6.   A summary of 8086 addressing mode.

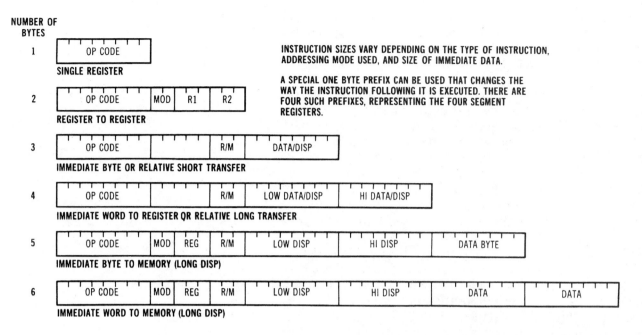

1 OP CODE

SINGLE REGISTER

INSTRUCTION SIZES VARY DEPENDING ON THE TYPE OF INSTRUCTION, ADDRESSING MODE USED, AND SIZE OF IMMEDIATE DATA.

A SPECIAL ONE BYTE PREFIX CAN BE USED THAT CHANGES THE WAY THE INSTRUCTION FOLLOWING IT IS EXECUTED. THERE ARE FOUR SUCH PREFIXES, REPRESENTING THE FOUR SEGMENT REGISTERS.

2 OP CODE | MOD | R1 | R2

REGISTER TO REGISTER

3 OP CODE | | R/M | DATA/DISP

IMMEDIATE BYTE OR RELATIVE SHORT TRANSFER

4 OP CODE | | R/M | LOW DATA/DISP | HI DATA/DISP

IMMEDIATE WORD TO REGISTER QR RELATIVE LONG TRANSFER

5 OP CODE | MOD | REG | R/M | LOW DISP | HI DISP | DATA BYTE

IMMEDIATE BYTE TO MEMORY (LONG DISP)

6 OP CODE | MOD | REG | R/M | LOW DISP | HI DISP | DATA | DATA

IMMEDIATE WORD TO MEMORY (LONG DISP)

Fig. 12-7. The 8086/8088 instruction formats.

$$0 \times 0 = 0$$
$$0 \times 1 = 0$$
$$1 \times 0 = 0$$
$$1 \times 1 = 1$$

EXAMPLE:

$$
\begin{array}{r}
1011 = 11 \\
\times 1010 = \times 10 \\
\hline
0000 \\
1011 \\
0000 \\
1011 \\
\hline
1101110 = 110
\end{array}
$$

Fig. 12-8. Rules for binary multiplication.

the op code, the instruction can contain either one or two bytes of data and either one or two bytes of displacement. Recall that the displacement is an address that is added to the base or index registers to form the offset part of the address. The 2-bit MOD, 3-bit REG, and 3-bit R/M portions of the instruction codes are sets of bits that designate mode, register, and register/memory selection. The MOD and R/M bits select the addressing mode, while the REG field selects which register will be used in the computation.

Six basic types of instructions are used in the 8086/8088. These are data transfer, arithmetic, logic, string manipulation, control transfer, and processor control. It would be impossible to cover the approximately 300 different instructions available in this book. However, to show you the power of 16-bit microprocessors over 8-bit devices, we will demonstrate some instructions that are unique to 16-bit processors.

An example of a 16-bit arithmetic instruction in the 8086/8088 is a multiply operation implemented with the MUL instruction. Both 8- and 16-bit words may be multiplied. The rules for binary multiplication are given in Figure 12-8. The product in this case is 110 decimal. The product could be as long as twice the length of the multiplier and multiplicand. For example, multiplying two 8-bit numbers results in a 16-bit product. In an 8-bit multiply, the content of accumulator AL is multiplied by the source operand, which is in another register or in memory. The address portion of the MUL instruction accesses the source. The product, which can be up to 16 bits in length,

(continued next page)

is stored in the AX (AL+AH) register. In 16-bit operations, the contents of accumulator AX is multiplied by a 16-bit source. The product, up to 32 bits long, is stored in registers AX and DX.

Another powerful 16-bit CPU instruction is MOVS or move string. This instruction is used to transfer a string of bytes or 16-bit words, from one area of RAM to another. The source string is addressed by the source index register SI and the destination of the string is specified by the destination index register DI. The MOVS instruction increments SI and DI as the data is transferred from one block of RAM to another. This is a good example of a CISC microprocessor instruction. Such an operation would have to be programmed in a RISC microprocessor.

The number of bytes or words to be moved is determined buy the quantity in the CX register. The repeat REP instruction is used in conjunction with MOVS to move a block of data. The MOVS instruction moves one byte or word each time it is executed. The REP instruction causes the operation to be repeated the number of times specified by the content of the CX register. The CX register is 16 bits wide and thus can specify up to a 65,536 byte/word block transfer.

The 8086 is which type of processor?

    a. RISC
    b. CISC

---

12   (*b.* CISC)   The input/output capabilities of the 8086/8088 are basically similar to those of other microprocessors. Programmed input/output operations are carried out by the special data transfer instructions. Interrupt and direct memory access operations can also be performed.

The 8086/8088 is capable of addressing 64K 8-bit or 32K 16-bit I/O ports. It is important to point out that these 64K I/O ports are addressed separately from the memory locations.

Although either the 8086 or the 8088 can transfer the data in 1-byte increments, recall that the 8088 has only an 8-bit data bus. Therefore, it is not as capable of transferring 16-bit-wide words as the 8086 is.

The 8086/8088 microprocessor can also use memory-mapped I/O. In memory-mapped I/O, the reserved I/O ports are not used. Instead, the input/output devices are simply attached to the data bus and given standard memory addresses. Each external device then appears to be a memory location. This permits the full power of the instruction set and the addressing modes to be used in input/output operations. All of those instructions that would normally be used to access data in memory or store it can be used in I/O operations with the various peripheral devices.

DMA operations can also be accommodated on the 8086/8088. Recall that in direct memory-access transfers CPU is not involved. The data is generally transferred directly between memory and an I/O device while the CPU remains idle. Special controller ICs, such as the Intel 8257 and 8237, are used to perform DMA operations. Control signals from the microprocessor tell when the address and data buses are free for the memory and I/O devices to exchange data.

A popular companion chip to the 8086 is the 8087, a math or numeric coprocessor. The 8087 works with the 8086 to provide full hardware implementation of floating-point and advanced mathematical operations. Basically, the 8087 is an extension of the 8086/8088 CPU. It adds registers and circuitry to perform a variety of special math operations beyond the usual add, subtract, multiply, and divide.

A special set of instructions allows the math capabilities of the 8087 to be used by a programmer. These instructions implement math operations on longer fixed-point data words. Some typical instructions include:

- add, subtract, multiply, and divide on 32- and 64-bit numbers
- square root
- compare 32- and 64-bit numbers
- long BCD word (packed decimal) operations
- 32- and 64-bit load and store instructions

It also implements floating-point operations on 32, 64, or 80-bit words. The 80-bit format is shown in Figure 12-9. Remember that floating-point numbers are the same as scientific notation whereby a number is multiplied by 10 raised to some power. The power of 10 multiplier sets the decimal point. Examples:

$$2.3 \times 10^8 = 230,000,000$$
$$7.5 \times 10^{-6} = .0000075$$

In floating-point binary, the number called the *significand* (also known as the mantissa) is multiplied by 2 raised to a power designated by the exponent. The exponent may be positive or negative as determined by the sign (S) bit.

Using the 80-bit format, the 8087 allows decimal values in the range 3.4×10^{-4932} to 1.2×10^{4932} with a 19-digit precision.

Some of the floating-point instructions include:

- load and store
- add, subtract, multiply, and divide
- square root
- logarithm
- tangent and arctangent

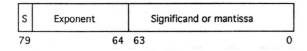

S	Exponent	Significand or mantissa
79	64	63 0

Fig. 12-9. 80-bit word format for Pentium floating-point data.

(continued next page)

Most early IBM PCs and compatibles using the 8088 or 8086 provided a socket on the main systems (mother) board for an 8087. If you wrote or ran software that incorporated the advanced math functions, you could install the 8087.

Although extra-long data word operations and floating-point operations could be programmed on the 8086/8088, such programmed operations consumed a great deal of time. By using the 8087 and its instructions, programs were shorter and considerably faster. In applications such as robotics, graphics, and image processing, calculations can be performed hundreds or even thousands of times faster. For example, a programmed square root subroutine might execute in 19,600 μS or 19.6 mS. Using the 8087 square root instruction hardware, a square root could be found in only 36 μS.

The 8087 math coprocessor can perform operations on long fixed-point words as well as floating-point operations.

   a. True
   b. False

---

13   (a. True)   Go to Frame 14.

The Intel X86 Series Microprocessors

---

14   Intel's 8086/8088 was superseded by the 80286 (or 286) microprocessor. It was introduced in IBM's PC AT computer in 1984. The 286 features a 16-bit data bus and a 24-bit address bus capable of addressing 16,777,216 (16M) words of memory. Early versions ran at clock speeds of 6, 8, 10, and 12.5 MHz. Later versions and clones from AMD ran at 16 and 20 MHz. The register organization and architecture of the 286 is similar to that of the 8086, so the 286 could run any software written for the 8086/8088. A numeric coprocessor called the 80287 was also available as a companion chip. A key feature of the 286 was its ability to implement virtual memory on an external disk drive. This allowed the 286 to generate virtual addresses for up to 1G (1,048,576) words of memory.

The 286, though widely used in IBM personal computers and the many compatible clones, was short-lived. Intel replaced it with the more powerful 80386 CPU. The 386 was Intel's first 32-bit MPU. It came in two versions, the 386DX and the 386SX. The 386DX had a 32-bit data bus and a 32-bit address bus, allowing it to address 4,294,967,296 (4G) words of memory. The 386SX had a 16-bit data bus and a 24-bit address bus, letting it address

16,777,216 (16M) of memory. The 386DX could run at clock speeds of 20, 25, 33, and 40 MHz. The 386SX could operate at 16 and 20 MHz. IBM made their own version of the 386 under license from Intel called the 386SL. Like the 386SX, it had a 16-bit data bus, but it had a 25-bit address bus, allowing it to address 33,554,432 (32M) words of memory. It ran at 20 or 25 MHz.

The heart of the 386 MPU was eight 32-bit general-purpose registers. The 386 also featured virtual memory and a pipeline with a 16-byte prefetch queue to increase computing speed. A math coprocessor was also available (387) if floating-point operations were needed. The 386 was fully compatible with older 8086 and 286 programs.

The first 32-bit design in the X86 series was the

    a. 186
    b. 286
    c. 386
    d. 486

---

15   (*c.* 386)   The 386 also had a relatively short life. Intel quickly replaced it with the 80486, or 486. The 486DX was a single chip containing a 386 MPU core to which was added an 8K on-chip level-1 cache, the floating-point coprocessor, and memory management circuits. The memory management circuits provide a flexible way to address the system RAM, ROM, cache, and virtual memory and to provide a way to expand RAM in logical increments. The 486DX had 32-bit address and data buses and ran at clock speeds of 25, 33, 50, 66, 75, and 100 MHz. Various 486 clones pushed 486 clock speeds to 120 MHz. A less powerful version called the 486SX had no math coprocessor, but one could be added (487). Clock speeds were 16, 20, and 25 MHz.

The 486 had a good run in personal computers, but its life also was cut short by Intel's introduction of the Pentium MPU. Intel dropped the X86 designation and instead of using the name 586, it coined the term *Pentium* for this next generations of CPUs.

The design of the 486 is similar to the 386, but the 486 has which of the following features?

    a. built-in cache
    b. higher clock speeds
    c. built-in floating point
    d. all of the above

---

16   (*d.* all of the above)   Go to Frame 17.

The Pentium Microprocessor

17 Although the Pentium microprocessor is a successor to the X86 series, it takes a giant leap beyond these earlier MPUs in computing power. First, the Pentium is a 64-bit microprocessor. It has 64-bit internal registers and 64-bit internal and external data buses. Like the other X86 models, it too has a 32-bit address bus that allows it to directly address 4G of memory. It is backward code compatible with earlier X86 models, thus preserving all of the previously written software.

The Pentium also uses internal cache, pipelining, and small geometry transistors (0.25 and 0.35 μ) to achieve extremely high speeds. Like previous Intel MPUs, Pentium is capable of pipelining instructions to increase speed. Pipelining implements instruction prefetch, which speeds overall execution. Although the 8086 had a 6-byte prefetch queue and the 486 had a 32-byte prefetch queue, the Pentium has two 64-byte prefetch queues. But unlike its predecessors it also features *superscaler architecture.* Superscaler means that the MPU has two CPUs that can execute two instructions simultaneously. Prefetched instructions can be executed alternately by the two CPUs, making computing even faster.

Figure 12-10 shows a simplified block diagram of the Pentium architecture. It has separate 8K instruction and data caches that bring data in and send data out to the external buses through the bus interface unit. These caches feed two integer execution units. An integer execution unit is essentially just complete CPU for processing fixed-point data. Each has its own 64-byte prefetch pipeline. This is the superscaler part of the Pentium. Note also that a single floating-point execution unit is also on-chip.

As the Pentium has evolved, its clock speed has also steadily increased. The earliest versions of the Pentium ran at 60 and 66 MHz clock speeds. Later 75, 90, 100, 133, 166, and 200 MHz versions emerged.

A more recent version of the Pentium is the MMX model. MMX stands for multimedia extensions. MMX Pentiums include 57 additional instructions for processing graphics, video, and audio thereby permitting higher speed handling multimedia applications such as games, TV, graphics on the Internet, and high-fidelity stereo digital audio. The MMX version also has larger (16K) data and instruction caches.

The MMX version of the Pentium is optimized for

    a. multimedia operations
    b. larger memory
    c. improved memory management
    d. memory-mapped I/O

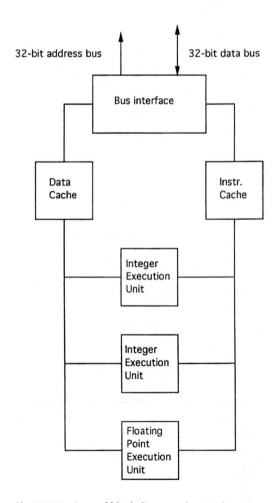

Fig. 12-10. General block diagram of an Intel Pentium microprocessor. Each integer execution unit is a 64-bit ALU.

18 (*a.* multimedia operations) Another more recent version of the Pentium is the Pentium Pro. The Pentium Pro is essentially a Pentium with an L2 cache built in. The on-chip 8K data and instruction caches are L1 caches. The L2 cache, unlike the L1 cache, is not on the Pentium chip itself. Instead it is on a separate chip. It is available in 256K or 512K word versions. The two chips are mounted on a common base (substrate), wired together, and then packaged into a single housing that looks like any other Pentium. The Pentium Pro is known as a hybrid IC.

The most recent version of the Pentium is the Pentium II. It is essentially a Pentium Pro with improvements. First, it uses the latest and smallest CMOS transistors, permitting even higher clock speeds. Versions with clock rates of 233, 266, 300, 333, 400, and 450 MHz are available as this is written, with even higher speed units expected in the near future. The Pentium II also has a larger 36-bit address bus that allows it to directly address 64 GB of memory. This is useful in advanced applications in network servers and engineering workstations.

The most controversial feature of the Pentium is its packaging. It abandons the traditional pin grid array (PGA) type of housing for a new package that is a plug-in plastic cartridge. Called a Socket 1 SEC cartridge, it features a Pentium Pro core CPU along with the 512K L2 cache chip on an internal substrate with a single-edge connector (SEC). The cartridge is designed to plug in to a special patented socket mounted on the computer's mother (systems) printed circuit board. The bus structure has been changed from that of other Pentiums to improve performance and to prevent competitors from copying the design.

Despite the Pentium's short life (it was introduced in 1993), it has undergone significant changes and improvements. In fact, the changes have occurred so fast that personal computer manufacturers have a difficult time keeping up with them. The minute a PC is announced, it becomes obsolete. Within six months to a year, the manufacturer will discontinue it for a newer model using the latest version of the Pentium.

Over the years several competitors of Intel have introduced Pentium clones. These independently designed processors are code-compatible with the Pentium and offer comparable or even improved performance at lower cost. Many computer manufacturers use the chips to make lower-cost PC models.

Commonly available clones are the Advanced Micro Devices (AMD) K5 and Cyrix 5x86 (Cyrix is a division of National Semiconductor) that duplicate the performance and operation of the original or Classic Pentium. The AMD K6 and Cyrix 6x86 have features and performance similar

(continued next page)

to the Pentium Pro. All of these processors are fully compatible but cost less than the Intel equivalent.

The K6 is made by

    a. Intel
    b. Cyrix
    c. AMD
    d. National Semiconductor

---

19   (*c.* AMD)   Go to Frame 20.

The Motorola 680xx Series Microprocessors

---

20   Another popular microprocessor family is the Motorola 68000. Next to the 8086/8088 series, the 68000 is probably the second most widely used 16-bit CPU. The 68000 is far more powerful than the 8086 and is probably more comparable to the Intel 286. The 68000 and its successors were used in the highly successful Apple Macintosh computers.

Refer to Figure 12-11. The 68000 has 17 general-purpose 32-bit registers plus a program counter and a status register. Whereas the registers in the 8086 are 16 bits in length, the 68000 registers all are 32 bits in length. Eight of these registers are designated as data registers, which are used to hold the various operands used in computing. Each of these registers may be accessed in 1-, 2-, or 4-byte segments, meaning that both 8-bit and 16-bit operations can be performed, as well as full 32-bit operations.

Seven of the registers in the 68000 are designated as address registers. These are all 32 bits long but can be accessed in 16-bit segments, if desired. These registers are used in various combinations to form memory addresses that access instructions or data. The 68000 also has two stack pointers, the user's stack pointer and the supervisory stack pointer. Although two stack pointers exist, only one is used at a time depending on the mode of operation.

Finally, there is a 32-bit program counter. Like program counters in 8086/8088 microprocessors, this PC points to the instruction to be executed. Although this counter has 32 bits, only 24 bits are used for generating external addresses. The address bus in the 68000 is 24 bits wide. This gives the 68000 the ability to direct address up to 16 megabytes (16,777,216 bytes) of memory.

The CPU also has a status register that contains the individual flag bits. The flags in the status register are set by the various CPU operations and, in some cases, by special instructions. The status bits are also monitored by various instructions to test various conditions that occur as a result of CPU operations.

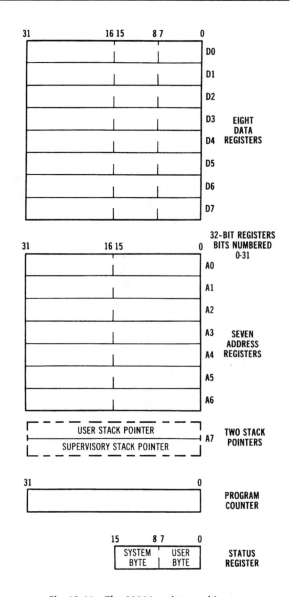

Fig. 12-11. The 68000 register architecture.

The 68000 has a total of 14 different operand addressing modes. The address of an operand is generally formed in one of the address registers. These registers can be used in a variety of ways, including indexed addressing. The final 24-bit address to locate an operand in memory comes directly from an address register, from the sum of two or more address registers, or from the sum of an address register and a displacement address that is part of an instruction.

The instructions for the 68000 vary in length from 16 bits to 64 bits. The op code, or op word as it is called in the 68000, is 16 bits long. Added to the basic op code that defines the operation are additional 16-bit segments containing an immediate operand, a source address, and a destination address.

The 68000 has a variety of different types of instructions. These include data movement, arithmetic, logical, shift and rotate, bit manipulation, BCD, program control, link and unlink, and system control instructions. Like most other 16-bit microprocessors, the 68000 features multiply and divide and other powerful instructions for greatly speeding up and simplifying various processing operations.

Note that the 68000 has no special I/O instructions. All input/output devices are treated as memory locations. By using memory mapped I/O, practically the entire instruction set is made available to the programmer for performing data transfers to and from external devices.

The 68000 also has a companion math coprocessor chip called the 68881. Other support chips provide memory management, DMA, and other commonly used functions.

The performance and features of the 68000 are closest to those of the Intel

a. 8088
b. 286
c. 386
d. 486

---

21 (b. 286)   Over the years, Motorola introduced more powerful successors to the 68000. These include the 68020, 68030, 68040, and 68060. The 030 and 040 are similar in performance and features to the Intel 386 and 486 respectively. These MPUs are used in various Apple Macintosh models. The most powerful 060 version is similar to the Pentium in performance, but was never used in personal computers. Instead it is used in special embedded-controller applications.

Table 12-2 summarizes the features of the later 680xx models.

(continued next page)

Table 12-2. Motorola 68020, 68030, and 68040 Features

Processor	Data Word	Data Bus	Address Bus	Other Features
68020	32-bits	16-bits	24-bits	256 word instruction cache.
68030	32-bits	32-bits	32-bits	256 word instruction and data caches. Built-in memory management with virtual memory.
68040	32-bits	32-bits	32-bits	8K word instruction and data caches. Built-in floating point coprocessor and memory management.

The 68030 and 68040 MPUs were most widely used in the

a. Intel computers
b. Motorola computers
c. IBM PC-compatibles
d. Apple Macintosh computers

22 (*d.* Apple Macintosh computers) Go to Frame 23.

The Power PC

23 The Power PC was a joint design effort of Motorola, IBM, and Apple. The end product was a series of 32-bit RISC processors. Apple uses various versions of this processor in its Power Macintosh personal computers. IBM also uses some versions in its RS6000 workstations and in various peripheral devices.

The basic features of the Power PC are a RISC architecture with 32-bit data and address buses, 32 64-bit general-purpose registers, and an instruction repertoire with less than 100 instructions. Built-in floating-point capability is standard in all models. The result is an amazingly fast processor that more than holds its own against the Pentium.

The first version, the 601, was introduced in 1993. It ran at 60 MHz, but later versions ran at clock frequencies up to 120 MHz. It also has an internal 32K word cache memory for both data and instructions. The 601 was later revised, updated and improved, resulting in the 603. Its organization is similar, but smaller geometry CMOS circuits boosted speeds up to and exceeding 350 MHz clocks. The 603e version was designed for lower operating voltage (3.3 vs. 5 volts) to greatly reduce the power consumption. Other power management features make the processor suitable for battery operation. This processor is used in Apple's PowerBook portable Macintosh.

The 604 version is similar architecturally to the 603, but does not have the power management features. Like the Pentium, it features two integer processors, so it is considerably faster because of the parallel superscalar operation that is possible. The 604 also has separate 16K data and instruction caches. The 604 also uses smaller transistors, so it is considerably faster. Versions with clock speeds of 200, 250, 300, 350, 400, and 450 MHz are available.

The most recent version of the Power PC is the G3 series for third generation. The G3 is a revised and updated design similar to the 604. Its smaller geometry circuits let it run at speeds up to 400 MHz. It buses are redesigned for higher speed and larger, faster cache memories are available. Versions are available with up to 1 M word of internal L1 cache, boosting performance further. Future versions are expected to feature a separate L2 cache built into the package, like the Pentium Pro.

Which of the following is not one of the features of a Power PC?

    a. built-in cache
    b. CISC design
    c. internal floating-point unit
    d. 32-bit data and address buses

---

24   (*b.* CISC design)   The Power PC is a RISC microprocessor.

Answer the Self-Test Review Questions on the diskette before going on to the next unit.

Personal Computers

Introduction

1   A personal computer is a general-purpose, desktop digital computer designed around a specific microprocessor. Personal computers were first developed back in 1975 after the first practical 8-bit microprocessors were introduced. Most of these early computers were primitive in that they came in kit form and had minimal peripheral (input/output) and storage devices. In the late 1970s, practical personal computers came along, such as the Apple II, Radio Shack TRS80, and models by manufacturers such as Commodore, Heathkit, and others. The practicality of the computers increased as more useful software was written for them. Hobbyists, engineers, and other persons with a technical background in building and programing computers owned these early PCs.

(continued next page)

Perhaps the most significant developments in personal computers were by IBM. In 1981, IBM introduced its first personal computer. It was an immediate success, not only with hobbyists and technically literate users but also with general business and industry. IBM produced a neat package of hardware and provided a considerable amount of usable software, including word processing programs, spreadsheets, databases, and graphics.

The IBM was so successful, in fact, that many new companies started developing clones or software-compatible computers. Because the IBM PC was made from standard components and was not patentable, others were free to copy it using existing components such as the Intel 8088 microprocessor, memory chips, disk drives, and video monitors. These compatible clone PCs were also enormously popular and further increased the use of personal computers at a staggering rate. The IBM PC was also so successful that it virtually eclipsed all other models, driving most of them into bankruptcy or into specialty niches.

In 1984, Apple introduced its successful Macintosh series of computers. Although never as popular as the IBM PC and its clones, it nevertheless carved out a huge market segment for itself. Many of Apple's features were patented and, therefore, could not be copied by others unless licensed by Apple. Apple chose to keep the entire Macintosh market to itself rather than permit cloning. This resulted in a more limited marketplace for their computers despite their success.

Over the years, the IBM PC and its many clones continued to change and improve as new microprocessors, memory chips, disk drives, and other advanced technology was introduced. Today, development continues at a rapid pace with a new generation of PCs being introduced almost annually.

Today most personal computers, regardless of the manufacturer, are the same. They all use the latest Intel microprocessor and a version of the Windows operating system from Microsoft. In fact, these personal computers have come to be known as *Wintel computers,* which is short for Windows–Intel computers. The variation from one model PC to another is so small that it is difficult to decide which one to buy. Most individuals buy their computer based upon a particular brand name, such as Compaq or Dell or IBM, and by price.

As for Apple, it too followed a path of continuous upgrading in microprocessor and computing power and software. However, because of marketing mistakes and the shear dominance of Intel computer business, Apple holds only a tiny share of the personal computer business (less than 5 percent as this is written). Today over 95 percent of the personal computers in use and being sold are some version of the basic Wintel design. The purpose of this chapter is to discuss the organization, operation, and features of this type of personal computer.

What is the name given to personal computers using Intel microprocessors and the Microsoft Windows operating system?

    a. Apple Macintosh
    b. IBM PC
    c. Wintel PC
    d. clones

---

2   (c. Wintel PC)   As you read this chapter, keep one very important point in mind: The information given here is current as of this writing. But given the rapid developments in the PC field, it may soon be more history than fact. Nevertheless, the fundamentals remain the same. The information provided here is more than adequate to acquaint you with the organization and operation of the modern PC and its successors.

Go to Frame 3.

PC Organization

---

3   All personal computers are virtually the same in that they consist of the following basic components:

    a. microprocessor (Intel Pentium or compatible, such as AMD K6)
    b. RAM, typically 16 MB, 32 MB, or more
    c. ROM
    d. Hard-disk drive (HDD) to store the operating system, application software, and files
    e. Floppy-disk drive (FDD) for diskettes
    f. CD-ROM drive
    g. modem or network interface card (NIC)
    h. video monitor
    i. keyboard
    j. mouse
    k. printer

A general block diagram of an IBM PC-compatible or clone computer is shown in Figure 13-1. You should immediately recognize that the personal computer is made of the same basic elements found in any computer, namely the CPU, memory, and I/O circuits. Also shown here are those peripheral devices, such as input/output devices and mass storage media, normally associated with a personal computer. In the typical PC, the microprocessor is one version of the Pentium or a compatible version, such as the

(continued next page)

AMD K6 or the Cyrix 6x86. These processors have 32-bit address and data buses to which all other peripheral devices and circuits are connected. In Figure 13-1, the narrow bus is the data bus. The control bus is not shown for simplicity. The buses connect to interface circuits that, in turn, connect the processor to the peripheral device itself.

Most of the circuitry in a personal computer is packaged together on a single large printed circuit board known as the *systems board* or *motherboard*. The motherboard contains the microprocessor, all RAM and ROM, and some I/O circuits. The motherboard also contains the buses over which data is transferred. In most modern personal computers there are two levels of buses, the local bus directly from the microprocessor and another bus designed to accept plug-in cards for slower peripheral devices. Those I/O circuits or interfaces not on the motherboard are plugged into sockets on one of the buses. Those interfaces generally *not* on the motherboard are designated within the shaded boxes in Figure 13-1. Now let's take a look at each of the major components shown in Figure 13-1.

Microprocessors

The heart of all computers, of course, is the CPU, in this case an Intel microprocessor. The majority (about 85 percent) of all modern personal computers have an Intel microprocessor. But over the years both AMD and Cyrix have captured some of the personal computer manufacturer's attention. The AMD K6 and the Cyrix 6X686 microprocessors are virtually 100 percent software-compatible with the Intel Pentium microprocessors and sell at a lower cost. A personal computer using these microprocessors will also run the Microsoft Windows operating systems and all of the software written for the Wintel computers. We have already discussed the Intel microprocessor in the previous chapter so we will not repeat that information here. Just remember that when selecting a personal computer, the specific model of the microprocessor will determine the performance, features, and future expansion of the PC. Personal computer models are available with the basic or classic Pentium, the Pentium with MMX, the Pentium Pro, or the Pentium II. MMX refers to extra on-chip hardware that implements numerous instructions for dealing with sound and video. Each year, PCs are used more and more with video, graphics, and sound. The MMX instructions speed up and simplify the writing of software for multimedia applications.

Each level of Pentium microprocessor is also available in several clock speed ranges. The older Pentiums have speeds of 60, 66, 75, 90, 100, 133, and 166 MHz while most of the newer models have clock speeds of 200, 233,

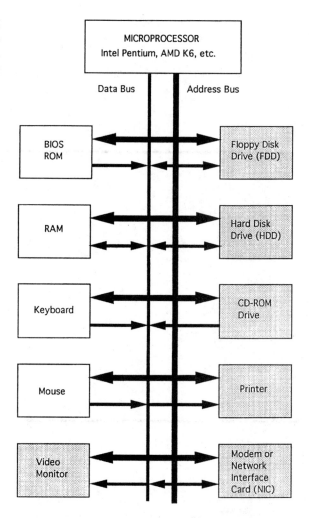

Fig. 13-1. General block diagram of the most common modern "Wintel" personal computer. (The shaded areas are peripheral devices.)

266, 300, 333, 400, and 450 MHz. Even faster speeds are expected in the coming years, as are different versions of the Pentium.

The microprocessor is usually plugged into a socket that is mounted directly on the printed circuit motherboard. The older Pentiums used a 296-pin PGA (pin grid array) package called Socket 5 while the Pentium Pros use a 387-pin PGA (pin grid array) package called Socket 7. The AMD K6 and the Cyrix microprocessors use the Socket 7 mounting arrangement. Most Pentiums or their clones also include a small fan for cooling or a heat sink to dissipate the enormous amount of heat generated.

The Pentium II is packaged in a completely different way. The Pentium II chip is mounted on a small printed circuit board along with a single L2 cache memory chip. The entire board is then enclosed in a small plastic module. An edge connector extension from the PC board is used for mounting. This package is called Socket 1 SEC (single-edge connector). The Socket 1 module plugs into a matching connector on the PC motherboard. This arrangement has two basic advantages. First, it provides a higher-speed bus connection to the L2 cache that greatly speeds up memory accesses. Second, it is a patentable feature that cannot be legally copied by AMD or Cyrix, thus minimizing CPU competition.

Which microprocessor model is *not* installed directly on the motherboard?

    a. Pentium Pro
    b. Pentium MMX
    c. AMD K6
    d. Pentium II

---

4   (*d.* Pentium II)   Go to Frame 5.

PC Buses

5   Remember that all microprocessors have three basic buses: the data bus, the address bus, and the control bus. The address and data buses are each 32-bits wide with a Pentium, as shown in Figure 13-1. (The control bus is not shown to simplify the diagram.) These buses connect the CPU to the various memory and I/O circuits. In some cases, the connection is made directly to selected circuits by lines on the motherboard. However, all personal computers have sockets provided to make bus connections to

(continued next page)

auxiliary I/O circuits and peripheral devices on plug-in cards.

The microprocessor bus is generally referred to as the *local bus*. It runs at the speed of the microprocessor but slows down immediately whenever you connect it to long lines on the motherboard, sockets, and other circuits. To minimize this problem, some intermediate buses have been developed to isolate the local bus from the connecting lines, sockets, and circuits to preserve the high-speed data transfer capability. These are sometimes referred to as *mezzanine buses*. The VL-bus developed by the Video Electronics Standards Association (VESA) to speed up video transfers in 486 systems is an example of a mezzanine bus. Today, the Peripheral Component Interconnect (PCI) bus is by far the most widely used mezzanine bus in Pentium PCs. The AGP bus is another.

Over the years of PC development, several different buses have been created and used. Table 13-1 lists and explains each of these buses.

Table 13-1. Personal Computer Buses

Name	Data Speed	Comments
PC bus	8-bits 8 MBps	Original bus used with 8088 CPU
Industry Standard Architecture (ISA)	16-bits 16 MBps	Bus used in AT model PCs using 286 CPU
Extended Industry Standard Architecture (EISA)	32-bits 32 MBps	Used primarily in some servers. Never popular.
Micro Channel Architecture (MCA)	16-bits 40 MBps	Proprietary IBM bus never widely adopted
VL-bus	32-bits 120 MBps	Local bus used for video interfaces. Obsolete.
Peripheral Component Interconnect (PCI)	32-bits 33/66 MBps	Now used in all PCs to speed up data transfers. Some versions of the PCI bus run at 100 MHz.
Accelerated Graphics Port (AGP)	32-bits 66 MBps	High-speed video bus for Pentium II systems.
PC card	16-bits 16 MBps	Created to use with laptop/notebook computers
CardBus	32-bits 33 MBps	Larger, faster version of PC card bus

Note: MBps = megabytes per second

The ISA bus, sometimes still called the AT bus, has been used for years. It first appeared in IBM's AT PC with a 286 processor. This bus continued to be used even as PC manufacturers adopted the later 386, 486, and Pentium processors. Only later, with the higher speed Pentium processors, did manufacturers find an alternative. The IBM MCA bus and EISA buses were attempts to create faster buses, but neither of these became popular. Today the PCI bus developed by Intel has been adopted by most PC makers as the bus of choice for the future. Most PCs include the PCI bus, but many continue to include the ISA bus because some of the slower peripherals still use it, and its presence in a PC lets owners continue to use some of their older plug-in accessories that use the bus.

The PC Card bus was a standard developed by the Personal Computer and Memory Card International Association for notebook computers. It was originally called the PCMCIA bus. This 68-pin bus accepts a thin interface card about the size of a thick credit card. The card contains either additional memory or an interface to some external device. Most PC cards contain interfaces for modems, network interfaces, or external disk drives. The CardBus is a larger higher speed version of the PC Card bus.

Which bus is still widely used in PCs?

    a. ISA
    b. EISA
    c. MCA
    d. VL-bus

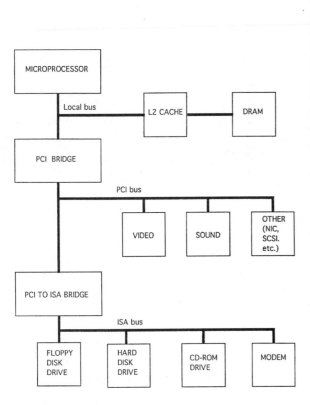

Fig. 13-2. Hierarchy of buses in a modern PC.

6 (*a.* ISA bus)   The primary characteristic of any bus is how fast data can be transferred. Since a bus is many parallel conductors that run over a distance, each bus line has resistance, inductance, and capacitance to ground. Each bus line acts like a transmission line, which has low pass filter characteristics. The longer the bus lines, the more pronounced the filtering effect. The effect of the bus is to round off, distort, and delay the binary signals. This greatly minimizes the speed of signals that the bus can accommodate and the speed of data transmission. The ISA and EISA buses have a maximum speed of 8 MHz. Since a 16-bit bus can carry two bytes and a 32-bit bus can carry four bytes, this speed translates to 16 MBps and 32 MBps (megabytes per second) data transfer rates.

Most personal computers today use the PCI bus and the ISA bus in an arrangement like that shown in Figure 13-2. The Pentium processor talks to the L2 cache and DRAM over the super fast internal local bus. This bus is then connected to a PCI bridge IC that buffers the local bus and creates a PCI mezzanine bus to which high speed peripheral devices connect. The video interface connects to the PCI bus as does several other high-speed devices, such as sound and network interface cards or peripherals using fast versions of the SCSI bus. Up to four sockets are usually provided to attach such interfaces.

The PCI bus speed is 66 or 100 MBps in current personal computers. Newer designs in the next generation PCs will extend this to 133 MBps and beyond.

(continued next page)

One of the PCI bridge ports connects to another IC called a PCI to ISA bridge. This circuit creates a standard ISA bus that is still used for connecting slower peripherals such as disk drives and modems. Several bus sockets are provided. Eventually, the ISA bus will no longer be found in PCs, but for now it is still useful.

The preferred PC mezzanine bus is the

    a. VESA bus
    b. PCI bus
    c. ISA
    d. EISA

---

7   (*b.* PCI bus)   Go to Frame 8.

Memory

---

8   Personal computers contain a mix of memory circuits, including ROM, static RAM (SRAM), and dynamic RAM. Let's take a more detailed look at each.

Static RAM

Static RAM is used for cache memory. The L1 cache is usually inside the microprocessor itself. In older PCs using 386 processors, separate SRAM chips on the motherboard were used. Over the years the trend has been to put both level 1 and level 2 cache inside the microprocessor itself. By shortening the leads on the bus to the cache memory, much higher speeds can be achieved. In older personal computers using the 386 and 486 processors, L2 cache was actually installed on the motherboard as separate chips or in a separate plug-in module.

Dynamic RAM

The random access read/write memory in a personal computer is usually implemented with dynamic RAM chips. These chips are generally mounted on small printed circuit boards containing an edge connector. These RAM modules are usually referred to as SIMMs (Single Inline Memory Module). Most SIMMs contain many megabytes (MB) of memory. Older computers used 256kB, 1 MB, or 4 MB SIMMs. The most recent plug-in memory modules are called dual in-line memory modules or DIMMs. These taller units contain larger DRAM chips and up to 64 MB or 128 MB of storage each.

Special sockets are provided on the motherboard for the SIMMs or DIMMs. The older SIMMs had a 30-pin edge connector. The newer SIMMs have a 72-pin edge connector that is plugged into one of usually eight separate slots available. The newer DIMMs have a 168-pin edge connector. Most modern personal computers come with a minimum of 16 MB or 32 MB of RAM, although that is increasing each year as DRAM storage density increases and prices continue to drop. In addition, DRAM is easy to expand by simply adding additional SIMMs or DIMMs.

DRAM chips are installed on

    a. the motherboard.
    b. small printed circuit boards that plug into the PCI bus.
    c. small printed circuit boards that plug into special SIMM or DIMM sockets on the motherboard.
    d. a special RAM board that attaches to an I/O bus.

9
(*c.* small printed circuit boards that plug into special SIMM or DIMM sockets on the motherboard.)

There are several different types of DRAM with which you should be familiar. These are FPM, EDO, and SDRAM. Fast-paging mode (FPM) DRAM is the older standard type. The memory is organized into pages to speed up the addressing of data. Most of the access time is involved with addressing the desired page. Once the page is located, addressing times are shorter, thus obtaining faster access to instructions and data. Access times are usually 70 or 80 nanoseconds (nS).

Extended data out (EDO) DRAM uses a scheme of setting up the access to the next address in sequence while data is being read out from the previous access. This makes the access time even faster. A typical access speed is 60 nS. A faster version of EDO-DRAM is burst EDO or BEDO-DRAM. It improves access time further since it accesses sequential data words and does not require an address set-up for each word.

The newest type of DRAM is synchronous DRAM. SDRAM is like BEDO-DRAM but synchronizes itself with the CPU clock. This results in a further improvement in access time. Some of the newer SDRAM circuits on DIMMs have an access time as low as 5 to 10 nS, approaching the speed of many cache memories.

One final note about DRAM: Some DRAM SIMMs and DIMMs are available with or without a *parity bit.* Parity is an extra bit added to each byte of memory for the purpose of detecting read and write errors. If an error is detected, an interrupt is generated to tell the CPU and the operating system. This is a signal that some problem with the RAM chip exists. A 1 MB DRAM chip without parity is organized as 1 MB × 8 or 1,048,576 8-bit words (bytes), for a total of

(continued next page)

8,388,608 bits. If parity is used each word has one additional bit, or 9 bits. The DRAM then has a 1 MB × 9 organization. A DRAM with 4 MB of 32-bit words has a parity bit for each byte, or four parity bits in this case, since there are four bytes per 32-bits. The organization then is 4 MB × 36. Lower-cost PCs use DRAM without parity to save money. PCs with parity DRAM are a little more expensive, but reliability is improved and errors, few as they may be, are reported.

The fastest type of DRAM is

    a. SDRAM
    b. EDO
    c. FPM
    d. DIMM

---

10   (*a.* SDRAM)   Go to Frame 11.

BIOS ROM

---

11   All personal computers contain a read-only memory called the BIOS ROM. BIOS stands for basic input-output system. The BIOS ROM contains a number of programs that are used for booting up (starting) the computer. Most of the programs to be run by the computer are installed on the hard disk drive, but these cannot be accessed until the computer is turned on and operating. The programs in the BIOS ROM set up and enable the computer to begin operation. Most of the programs are input and output programs that allow the computer to speak to the keyboard, the mouse, the video monitor, and the disk drives. The BIOS ROM also contains a power-on self-test (POST) program that does an initial test of the microprocessor, RAM, and I/O devices. If all is well, the computer is ready to run and the BIOS programs turn the operation of the computer over to the operating system. On most modern computers, this operating system is the Microsoft Windows 95 or some earlier or later variation. Under the control of the operating system, the computer is ready to use. From there you can select the programs and applications you wish to run.

Most of the programs in the BIOS ROM are related to

    a. memory
    b. CPU test
    c. I/O
    d. the operating system

---

12   (*c.* I/O)   Go to Frame 13.

13 The remainder of the circuits in the PC are I/O interfaces. Most of these are on separate printed circuit boards that plug into the connectors on either the PCI or ISA buses. All I/O interfaces use interrupt I/O, and the high-speed data storage peripherals such as the disk drives use direct memory access (DMA). The various I/O interfaces usually found in a modern PC and their usual configuration follow.

1. *Keyboard.* Interface on the motherboard. Newer systems use a Universal Serial Bus (USB) port.
2. *Mouse.* Interface on the motherboard in newer PCs, but on a separate plug-in card in older PCs. Newer PCs use a USB port.
3. *Printer.* Interface on the motherboard in newer PCs but on a separate plug-in card in older PCs. Usually a parallel port, but RS-232 serial ports have also been used. On newer systems a USB port is used.
4. *Video interface.* Usually on a separate card that plugs into the PCI bus on the newer PCs but into the ISA bus on older PCs. The Accelerated Graphics Port (AGP) bus is available on newer Pentium II-based systems for video.
5. *Floppy disk controller.* On a separate card in older systems but within the floppy disk drive itself in newer systems. Older designs packaged the floppy and hard disk controllers on a common board.
6. *Hard disk controller.* On a separate card in older systems but within the fixed disk drive itself in newer systems.
7. *CD-ROM drive.* On a separate card in older systems but within the floppy disk drive itself in newer systems.
8. *Modem.* A separate plug-in card that plugs into to the ISA bus.
9. *Network interface card.* Sometimes on the motherboard in newer PCs, but also a plug-in card for either the ISA or the PCI buses.
10. *Sound.* A separate board on most PCs that plugs into the PCI bus.

Refer back to Figures 13-1 and 13-2 to see typical arrangements.

The I/O interface for most disk drives is contained in/on:

a. the motherboard
b. the disk drive itself
c. a special interface card for the ISA bus
d. a DIMM

14   (*b.* the disk drive itself)   Peripheral devices and their interfaces will be discussed in more detail in the next unit.

Go to Frame 14.

Power Supply

15   All PCs have a built-in power supply that receives the 120-volt 60 Hz AC from the power line and converts it into several DC voltages to operate all of the other devices in the computer. Most PCs have four DC outputs, +5 volts, –5 volts, +12 volts, and –12 volts. These voltages connect to the PC motherboard and the disk drives through plastic connectors called Berg connectors. Most of the logic circuits on the motherboard operate from the +5 volt supply. The microprocessor typically operates from +3.3 volts, which is derived from the +5 volt supply through a separate regulator on the motherboard. By using 3.3 volts on the CPU chip, its power consumption and heat dissipation is cut dramatically and its speed is increased. In the future, more and more logic circuits will use 3.3 volts to further increase speed and reduce power consumption.

The disk drives operate from +5 volts for the logic and +12 volts for the drive motors. All other plug-in boards commonly use +5 volts; however, some peripherals and interfaces may also use the –5- and –12-volt lines.

The power supply is easily identified in the PC because it is enclosed in its own sealed, ventilated steel case. It is installed at the rear of the computer with its cooling fan facing the rear. The AC cord plugs into a socket in the rear. Some power supplies also have a 110/220 volt switch on the back panel so that either 110/120 or 220/240 volt AC input may be used.

Power supplies are rated by the amount of power they can supply. Power is stated in watts. The smallest PCs with little expansion capacity may have a power supply with a rating of 100 to 150 watts. Most desktop PCs have a power supply that will put out 150 to 200 watts. If you have a PC packaged in a mini-tower, a larger power supply of 150 to 250 watts is normally included to handle the extra accessories that these larger PCs can hold. Larger PCs, particularly servers in a large, full-tower cabinet, have considerably extra room and plug-in slots for many additional disk drives or I/O interfaces. The power supply in such a machine can put out up to about 300 watts. If you plan to expand your machine with extra memory and peripherals, be sure to buy a PC with extra power capacity.

All PC power supplies are of the switching variety. A basic full-wave bridge rectifier converts the 120 VAC input to an unregulated DC that drives a high-frequency square wave power oscillator. The high frequency AC generated by this circuit is then fed to a transformer with multiple secondary windings. The AC voltages from these windings are rectified and then regulated to form the four basic DC output voltages. Peripheral devices that have their own power supplies are the video monitor and the printer.

The most commonly used DC power supply output is

    a. +3.3 volts
    b. +5 volts
    c. +12 volts
    d. −5 volts

---

16   (*b.* +5 volts)

Peripheral Equipment

Peripheral Equipment

1 A peripheral device is any circuit or piece of equipment that is separate from or external to a microcomputer. The microcomputer consists of the CPU, RAM and ROM, and the I/O circuits. Peripheral equipment consists of all other devices that are connected to the computer for input/output or storage.

(continued next page)

The two basic types of peripheral equipment used with computers are input/output devices and auxiliary memory. Input/output devices are units used to communicate with a computer. Humans do not inherently think, speak, or otherwise communicate in the binary language that the computer understands. Humans use decimal numbers and the letters of the alphabet to make up words, sentences, and larger blocks of text. Input peripheral units are used to convert the language and symbols of the human operator into binary data that the computer can manipulate. Output peripheral units convert the binary data of the computer into understandable human information. These units serve as a way to get data into and out of the computer. Typical input devices are the keyboard and mouse. The most common output devices are the video monitor and the printer.

The other major category of peripheral equipment used in personal computers is auxiliary memory. Also known as mass storage, these peripherals are used where a considerable amount of data and a variety of programs must be stored. PCs typically rely on their memory (RAM) to contain their programs and data. The PC is a general-purpose machine that can run many different programs. These programs are stored in some external and/or removable media. When the desired program is to be used, the user calls for them and the operating system loads them into main memory where they are executed. Many larger programs are run by bringing some of the program into the RAM and then accessing others parts stored on the hard drive when needed. The most common mass storage peripherals are the diskette or floppy-disk drive, the hard drive, the CD-ROM and DVD optical drives, and magnetic tape drives.

Although peripheral devices are generally considered to be separate units from the computer itself, in reality many peripherals are packaged together physically with the computer circuits. In a personal computer, the keyboard, mouse, and video monitor are physically separate from the computer "box," but the disk drives are contained within the computer enclosure.

The two main types of peripheral devices in a computer are

    a. input and output
    b. I/O and mass storage
    c. internal and external
    d. manual vs. automatic

---

2   (*b.* I/O and mass storage)   Go to Frame 3.

3 The two most common input devices for a personal computer are the keyboard and the mouse. Let's take a more detailed look at each of these devices.

Keyboard

The keyboard is a major subsystem of a personal computer. On desktop computers the keyboard is a separate unit connected to the system by a cable. Of course, in a notebook or laptop computer the keyboard is part of the main housing. Most keyboards look like a typewriter keyboard but with some additional keys. One of these is the Enter key that serves as a carriage return and line feed or as a signal to accept the input given. Other keys include the control key (Ctrl), the alternate key (Alt), and the delete key (Del). The shift key, as on a typewriter, permits upper- and lowercase characters to be created and alternate characters to be selected.

Most modern PC keyboards also include a row of function keys labeled F1 through F12. These are arranged horizontally above the regular keyboard characters. These special function keys permit software writers to define these keys as required by a particular application. To the right of the regular keyboard area is the numerical keypad. These ten keys are arranged in the standard calculator format so that numerical data can be entered faster and easier. Finally, a special group of "arrow" keys (left, right, up, and down) is provided to move the cursor around on the screen.

Which key is used to signal the software to accept data provided in previous keystrokes?

> *a.* Alt
> *b.* Ctrl
> *c.* Enter
> *d.* Shift

4 (*c.* Enter)   The individual keys in the keyboard are switches. A variety of different switch types have been used in keyboards. They can be standard mechanical contact electrical switches, capacitive sensors, or magnetic sensors. Regardless of the configuration, pressing a key creates a contact closure or its equivalent that is sensed by the keyboard circuitry to generate a binary code representing that character. In addition, the circuitry generates an interrupt to the main CPU indicating that a key has been pressed and that input service is required. All of these func-

(continued next page)

tions are carried out by the electronic circuitry in the keyboard and interface circuitry located on the motherboard.

Figure 14-1 shows a general block diagram of the keyboard circuitry. A single-chip 8-bit microcontroller handles all of the keyboard functions. The program stored in the embedded controller ROM is designed to send a strobe pulse sequentially on each of the thirteen vertical outputs in the keyboard matrix. At the same time, the microcontroller scans the eight horizontal input lines from the keyboard matrix looking for a switch closure. If any of the key switches is depressed, the strobe pulse will ultimately appear on the row associated with that key. The strobe pulse will be transferred through the switch to the column line and read into the microcontroller. A delay subroutine waits

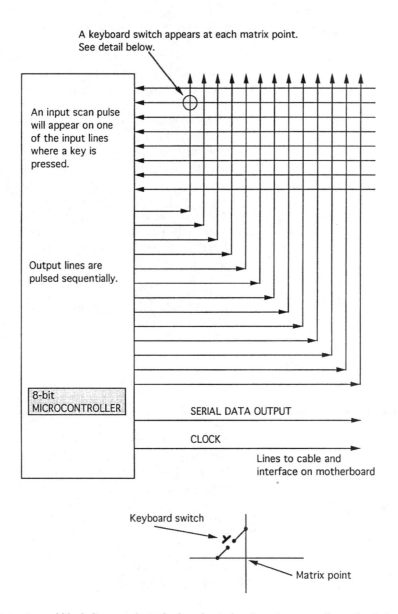

Fig. 14-1. General block diagram of a PC keyboard unit showing microcontroller and switch matrix.

a short period of time to avoid the contact bounce associated with switches.

The main component in a keyboard is a(n)

a. microcontroller
b. key switch
c. memory
d. interface circuit

---

5 (*a.* microcontroller)   The control program in the internal microcomputer translates the row and column information into an 8-bit Kscan code. The Kscan code is not the ASCII value associated with this key. It is a special code used inside the keyboard system. The scan code is then transmitted serially to the motherboard via a cable. This cable contains five connections, which include the positive supply voltage +5 volts, system ground, a reset line, the serial data, and a clock signal. The clock signal from the internal keyboard microcontroller ensures that the serial data transmitted has the correct synchronization with the keyboard interface on the motherboard. This cable terminates in what is called a DIN connector that plugs into the back of the PC systems unit. In most systems, another embedded microcontroller on the motherboard handles the keyboard interfacing.

The format of the Kscan code transmitted to the system unit is shown in Figure 14-2. The 8-bit Kscan code is transmitted within an 11-bit serial frame consisting of a binary 0 start bit, the 8-bit Kscan code that is sent LSB first, an odd parity bit, then a binary 1 stop bit.

The microcontroller on the system board receives the serial data and checks parity for transmission errors. It also translates the Kscan code into another 8-bit scan code. Finally, it generates an interrupt to the main processor. The generation of the interrupt causes the CPU to branch to a subroutine in the BIOS ROM. This subroutine then translates the scan code byte into the standard ASCII code. This code is then stored in a buffer area of RAM set aside for keyboard data. Finally, the operating system associated with the PC contains additional programs that will interpret and use the keyboard information.

How is the data from the keyboard transmitted to the PC?

a. parallel
b. serial

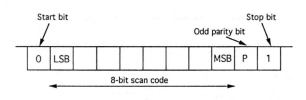

Serial data output from keyboard.

Fig. 14-2. Serial data output from the keyboard.

---

6   (*b.* serial)   Go to Frame 7.

Mouse

7   Most modern personal computers use a pointing device called a *mouse* to position a cursor on the screen for the purpose of identifying an object to be selected. The operating system with its graphical user interface (GUI) presents on the video screen a selection of icons, buttons, and menus that provide a means of choosing a desired program or operation. The mouse is a handheld device that is moved horizontally on the desk. Its motion creates movement of a cursor or pointer. The cursor is usually an arrow that is positioned over the desired icon, button, or menu item to select it. The mouse also includes two or three pushbutton switches that are used to signal the computer that a particular item is to be selected. The use of the mouse and graphical user interface software eliminates the need for users to keyboard in commands and other information that must be remembered.

The mouse is used to move

      a. text
      b. data
      c. windows
      d. the cursor

8   *(d.* the cursor)   Over the years a variety of mouse types have been designed. The most commonly used is the mechanical mouse that translates physical position into binary codes that are sent to the computer. Figure 14-3 shows the physical arrangement of the components in the mouse. A small ball about one inch in diameter with a rubberized coating appears on the bottom opening of the mouse. As the mouse is moved, the ball rotates. As it rotates, the ball comes in contact with two rollers that are mounted at right angles to one another. The rollers are connected to small disks that contain tiny holes or slots. These small wheels rotate inside an optical sensor unit consisting of a light-emitting diode (LED) and a phototransistor. As the mouse is moved, the ball rotates and turns the roller that, in turn, rotates the wheel between the LED and the phototransistor. The holes in the wheel cause the phototransistor to see intermittent pulses of light that it converts to binary pulses. Since the two rollers are at right angles to one another, two sets of pulses are generated, one that corresponds to the up/down movement of the mouse and the other to the left/right movement of the mouse. These pulses are fed to an integrated circuit that translates them into a serial data signal that is transmitted to the system motherboard via a cable.

A mouse can also have two or three pushbuttons that are also used for signaling the computer. The most often used is a left pushbutton. Once the cursor has been positioned

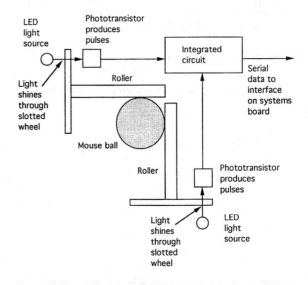

Fig. 14-3. Block diagram showing the mechanical and electronic features of a mouse.

over an icon on the screen, the left button is pressed, signaling the operating system that a choice of operation or software has been made. A right pushbutton is used for alternate operations. One type of mouse includes a center third pushbutton for vendor-defined operations. The two-button mouse is by far the most common.

There are two basic types of commonly used mouse interface connections. These are the serial mouse and the bus mouse. The *serial mouse* uses a standard RS-232 serial data format. The mouse is connected to one of the computer's RS-232 serial ports by way of a 9-pin D connector. Another type of mouse, called a *bus mouse,* transmits the serial data from the mouse to a special interface card plugged into the PC bus. The interface on the card converts the serial data into parallel data that is passed on to the system bus for interpretation by the system's software. Newer personal computers use the universal serial bus (USB) to connect the mouse to an interface on the main system board.

Regardless of the interface, the serial data from the mouse is converted into binary data that is interpreted by the BIOS subroutines and the operating system software. This, in turn, provides the data to the video system in the PC for generating the cursor and causing its motion.

How do you select a program or operation with a mouse?

    a. Move the cursor until it touches the icon.
    b. Move the cursor until it touches the icon and then press the left button.
    c. Move the cursor until it touches the icon and then press the right button.
    d. Move the icon until it overlays the cursor then click the left button.

9   (*b.* Move the cursor until it touches the icon and then press the left button.)   Go to Frame 10.

Output Devices

10   The two most common output devices used with personal computers are the video monitor and printer. Most modern personal computers also have provisions for a sound or audio output.

(continued next page)

The Video Subsystem

The video display system in a personal computer is by far the most complex part of any personal computer. It consists of two major components: the video monitor and the video controller. The video monitor is the display unit itself, while the video controller is the special interface card that converts binary data from the computer into the characters, graphics, and color to be displayed.

The video monitor is similar to a color television set but without the channel selecting circuits. It consists of a cathode ray tube (CRT), or picture tube, and the related electronic circuitry. The basic purpose of the CRT is to convert a scanning electron beam into light. Figure 14-4 shows a side view of a monochrome (black and white) CRT and how it produces light. An electron gun within the CRT generates a very narrow, focused beam of electrons that is accelerated toward the screen by thousands of volts. The inside of the screen is coated with phosphorous, a material that emits light whenever it is bombarded by electrons. A point of light is produced where the electron beam strikes the phosphorous.

What actually produces light on the face of a CRT?

   a. the filament in the electron gun
   b. ion bombardment of the screen
   c. electrons from the gun striking the phosphorous coating
   d. high voltage gas discharge

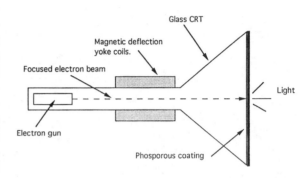

Fig. 14-4. Physical construction of a monochrome cathode ray tube (CRT).

11 (c. electrons from the gun striking the phosphorous coating) A magnetic coil called a *yoke* is placed around the neck of the CRT. Special scanning pulses are applied to produce magnetic fields that are used to deflect the electron beam. By controlling the pattern of the pulses to the magnetic coils, the electron beam can be made to scan across the screen at high speed. The deflection circuits sweep the beam across the screen in a sequence of horizontal lines that are displayed close together sequentially from top to bottom. The voltages applied to the yoke coils sweep the electron beam across the screen from left to right. The beam is then blanked off as it rapidly returns to the left to start another left-to-right scan slightly lower on the screen. Blanking the beam during this so-called retrace period prevents it from creating light on the screen during that time. Since the human eye is unable to follow high-speed scanning operations, what a person sees is a continuously lighted screen known as the *raster*.

To write information on the screen, the intensity of the electron beam is controlled. The simplest form of control is turning the beam off and on. This produces areas of black and white. By controlling the intensity of the electron beam, light levels between black and some high intensity

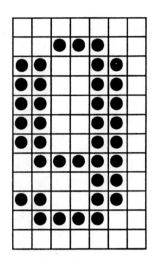

Fig. 14-5. A dot matrix video character.

white value can be achieved. Intermediate levels of light are interpreted by the eye as shades of gray.

Characters such as letters, numbers, punctuation marks, and special symbols are displayed by turning the beam off and on in a series of dots. Each character is formed as a sequence of on or off dots. These are called *dot matrix characters*. The dots are referred to as *picture elements* or *pixels*. Figure 14-5 shows a 7 × 12 matrix that is used to define all common characters. When the dots or pixels are close together, the character when viewed from a distance looks complete.

The entire video screen in a PC is usually defined in terms of the total number of pixels that can be displayed. The total number of pixels is equal to the total number of horizontal scan lines multiplied by the number of dots or pixels per line. A screen with 480 lines with 640 pixels per lines has a resolution of 480 × 640 or 307,200 pixels. The smaller the pixel, the finer the resolution and the better the display.

What is a pixel?

a. a colored dot
b. a single scan line
c. a type of CRT
d. a point of light

12 (*d.* A point of light)   A color CRT contains three electron guns that are deflected or swept simultaneously. These three narrowly focused electron beams are aligned to strike tiny red, blue, and green dots or rectangles on the inside of the screen. These three dots or rectangles form a set of color patterns that is repeated thousands of times adjacent to one another on the inside of the face of the CRT. The phosphors are designed to emit red, blue, or green light when they are struck by the appropriate electron beam. One of the basic principles of physics is that any color of light can be created by properly combining red, green, and blue light in the correct proportion. By varying the intensity of the red, green, and blue electron beams, any color pixel can be created on the screen.

A general block diagram of a color video monitor is shown in Figure 14-6. It consists of the CRT, the deflection circuitry for sweeping the beams horizontally and vertically, and the circuitry for controlling the intensity of the red, green, and blue electron guns. Other circuitry provides for horizontal and vertical positioning of the raster and overall video intensity.

The monitor receives signals from the video controller through a cable and a 15-pin D connector. The red, blue, and green video signals are analog voltage levels corresponding to a particular intensity of red, blue, or green light to be produced by the CRT. Special synchronizing or

(continued next page)

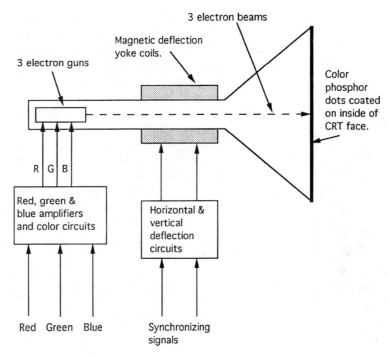

3 electron beams

Magnetic deflection yoke coils.

3 electron guns

Color phosphor dots coated on inside of CRT face.

R G B

Red, green & blue amplifiers and color circuits

Horizontal & vertical deflection circuits

Red   Green   Blue

Synchronizing signals

Input signals from video controller circuits.

Fig. 14-6. Block diagram of a color video monitor.

timing pulses control the horizontal scan rate and the vertical synchronizing rate. The typical horizontal scan rate is 31.5 kHz, which means that it takes approximately 31.75 microseconds for the beam to scan one horizontal line. The vertical refresh rate, or the rate at which the entire screen is scanned, is either 50, 60, or 72 Hz. A 72 Hz rate means that the entire screen is scanned 72 times in one second. Synchronizing signals are developed by the video controller in the computer.

Color is produced on the screen of a CRT by mixing light of

    a. red, blue, and yellow
    b. cyan, magenta, and yellow
    c. red, green, and blue
    d. black, white, and gold

13   (*c.* red, green, and blue)   The *video controller* or *video interface* consists of all of the electronic circuitry that translates the binary information about what to display into the signals that actually produce the output on the CRT. The intricacies of the video controller are far beyond the scope of this book. However, their primary purpose is to generate a series of dots or pixels and write them across the screen in sequential order. The intensity and color of each dot can be precisely controlled. Turning the red, green, and blue electron beams off and on or setting them to a particular intensity level gives the desired color and intensity of the pixel. Signals related to the synchronizing pulses are used to identify the precise location of each pixel on the screen.

Since the initial development of the personal computer, a wide variety of screen formats have been developed. The early versions were for monochrome display only and consisted primarily of dot matrix characters and simple line graphics. Over the years, different color formats have been developed. One of the most popular was the VGA or video graphics adapter format that produced 640 pixels on each of 480 lines (640 × 480). This produces a total of 307,200 pixels on the screen.

Today most personal computers come with an improved or super VGA (SVGA) controller. This controller can be set to display 600 lines of 800 pixels (800 × 600) or 1,024 pixels on each of 768 lines (1,024 × 768). Such high resolution produces very finely detailed characters and graphics.

Keep in mind that binary information is used to define the characteristics of each pixel. For example, an 8-bit binary number defines one of 256 voltage levels for each of the red, green, and blue electron guns. Each electron gun is able to produce 256 color levels, so the total number of colors that can be produced is 256 × 256 × 256 or 16,777,216. Although most personal computers do not use this many colors, the capability is there if an appropriate controller is used. Most personal computers allow you to select the color resolution. Typically, 256 different colors or shades are used in SVGA graphics.

The video controller on a PC is typically contained on a printed circuit board that is plugged into one of the bus slots. The key part of the video controller is a special random access memory known as *video RAM* (VRAM) where the binary input information defining each pixel is stored. These defining characters are read out at high speed and fed to digital-to-analog converters (DAC) that generate the red, green, and blue analog signals that are fed to the monitor.

Which of the following is *not* usually part of the video controller?

    a. video RAM
    b. DACs for each color
    c. sweep control circuits
    d. microcontroller

---

14   (*d.* microcontroller)   Go to Frame 15.

The Printer

15 A printer is the next most common PC output peripheral. The majority of personal computer applications for business, education, and home use require some type of hard copy paper output. A printer translates the binary text and graphic information into printed form.

Many different printing technologies have been developed over the years. Today the most common printers used with personal computers are impact printers, ink jet printers, and laser printers.

Impact Printers

Impact printers transfer characters and graphics to paper by means of a print hammer. It contains the character to be printed and strikes an inked ribbon which, in turn strikes the paper. The most common form of impact printer is the dot matrix printer.

In the dot matrix method of printing, the characters are not fully formed. Instead, they are made up of a series of small dots that when printed close together resemble all of the various letters, numbers, and other characters. The dots are arranged in a standard matrix format so that virtually any character can be created.

The simplest form of matrix is the five by seven (5 × 7). Refer to Figure 14-7A. Because of the small number of total dots available, the shape of the character is somewhat of a compromise, but the characters are fully recognizable. By using a higher-resolution dot matrix, more completely formed characters can be printed, as shown in Figure 14-7B.

Dot matrix characters are printed by a special print head made up of a group of vertically positioned and closely spaced vertical wires. See Figure 14-8. The end of each print wire can independently strike the paper through the ribbon. The print head prints one vertical column of the character matrix at a time. The print head is stepped horizontally across the paper in small increments so that each vertical column of dots is printed. Although this method of printing may seem clumsy and slow, the overall result is actually quite good. Even though the character is printed one vertical column at a time, it is nevertheless extremely fast. Because the dots are very small and closely spaced, the characters are easily recognized.

Improvements in dot matrix printers have been made over the years so that now many more print wires are used to improve resolution. One common arrangement is to use two vertical rows of nine wires slightly offset from one another as shown in Figure 14-9. Twenty-four pin print heads with three vertical rows of eight print wires have also been developed. By firing the print wires in the correct sequence, tiny overlapping dots are produced, creating an

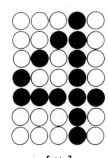

5 × 7

(A) A 5 × 7 pattern displaying the number 4.

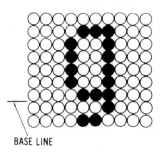

BASE LINE

10 × 9

(B) A 10 × 9 pattern displaying a lower-case g

Fig. 14-7. Typical dot matrix printing patterns.

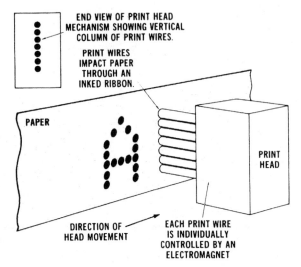

END VIEW OF PRINT HEAD MECHANISM SHOWING VERTICAL COLUMN OF PRINT WIRES.

PRINT WIRES IMPACT PAPER THROUGH AN INKED RIBBON.

PAPER

PRINT HEAD

DIRECTION OF HEAD MOVEMENT

EACH PRINT WIRE IS INDIVIDUALLY CONTROLLED BY AN ELECTROMAGNET

Fig. 14-8. Dot-matrix print head mechanism.

Fig. 14-9. 18-pin print head for NLQ output.

extremely high quality character. The resulting output text is usually referred to as near letter quality (NLQ).

Overlapping dots produces characters that are

a. perfect
b. a compromise
c. near letter quality
d. barely legible

16 (*c.* near letter quality)   The actual dot matrix character to be printed is usually stored as a bit map inside a read-only memory. To print a particular character, the ASCII code for that character is transmitted to the printer. The printer circuitry, typically an embedded microcontroller, looks up the desired dot matrix pattern in the ROM. Then the vertical column bit patterns are read out sequentially and fed to the print head to control the print wires.

It is possible for dot matrix printers to print several hundred characters per second. Further, different dot matrix patterns can be programmed into memory. This permits different types of print fonts to be used by simply plugging in a different ROM or downloading dot matrix characters into a memory from a computer.

Further, the dots can be used to create virtually any shape, meaning that dot matrix impact printers can also produce graphics. Lines and simple drawings can be easily reproduced. Limited use of color is also possible by changing the ink ribbon.

Like any other impact printer, dot matrix printers are also noisy. But their versatility makes them valuable. Perhaps their greatest use today is in printing multi-part forms. Because the print hammer impacts the paper through the ink, multiple copies can be produced with carbon paper or special duplicating paper forms.

Which of the following is the primary benefit of an impact printer?

a. highest quality print
b. high speed
c. easy to interface
d. good color usage

17   (*b.* high speed)   Go to Frame 18.

Ink Jet Printers

18 One of the most recent print technologies and one of the most popular today is ink jet printing. Ink jet printers are dot matrix printers that form characters by squirting tiny dots of ink at the paper in the correct format and sequence. Because the ink jets or dots are extremely small, very high resolution printing is achieved.

Ink jet printers are similar in their physical layout and electronic operation to dot matrix impact printers. A print head is moved horizontally across the page as the dots are formed. The print head in the case of an ink jet printer is a special cartridge containing ink and the mechanism for forming characters from tiny dots of ink that are squirted at the paper. Hewlett Packard developed the first practical, low-cost ink jet printers. They patented a piezoelectric print head mechanism that forms the ink into small dots and forces them out under pressure. Later, Canon patented an alternative technique that uses a print head, which forms a small dot of ink that is forced out by a small heat-produced bubble.

Regardless of the exact technique and mechanism of the print head, ink jet printers form characters with a minimum resolution of 300 dots per inch (dpi). Compared to dot matrix impact printers, this is far greater resolution. Most low-cost ink jet printers have a resolution of 300 or 360 dpi. Higher-priced ink jet printers can achieve resolutions up to 720 dpi.

In addition to high-resolution letter-quality printing, ink jet printers can also generate outstanding graphics. Further, by using multiple print heads with different color inks or special multi-ink print heads, high-quality color printing of both text, graphics, and images can be produced.

Although ink jet printers are still relatively slow, their high-quality printing more than makes up for the loss of speed. Another key advantage of the ink jet printer is that it is almost completely silent.

An ink jet printer is a dot matrix type.

    a. True
    b. False

19 (*a.* True)   Go to Frame 20.

Laser Printers

20 The highest quality printing and the highest print speeds are achieved with laser printers. A laser printer uses an electrophotographic printing technique like that used in Xerox machines or similar photocopiers.

Laser printers are page printers, that is, they print an entire page of information at a time. Unlike impact and ink jet printers that print one line at a time, laser printers form the entire image for one page and print that page before moving on. The job of forming the page is left to the software.

Laser printers are also dot matrix printers in that the images that they print are formed by extremely small dots of toner ink. The earliest laser printers used a 300 dpi format, but most modern laser printers use a 600 or 1200 dpi resolution, producing outstanding resolution and high-quality printing.

The laser printing method incorporates both optics and electrostatic techniques. The image to be printed on one page is transferred to a cylindrical drum in the form of an electrical charge. Electrically charged toner ink is then attracted to the areas of the drum to be printed. This image is then transferred to the paper and fused or hardened by the application of heat.

Figure 14-10 shows a simplified diagram of how a laser printer works. This drawing shows an end view of the drum, rollers, paper, and other components. The key component in a laser printer is the photosensitive cylindrical drum. The drum is coated with a material that is sensitive to light. Prior to painting the image on the drum, the drum is rotated and exposed to a high voltage charge applied by a corona wire. This fine wire placed near the drum has a very high negative voltage applied to it (several hundred volts). This places a very high negative charge on the entire drum surface. When the drum is exposed to light, it creates areas of positive and negative charges in the pattern of the characters or graphics to be printed. The information to be printed is formatted into a dot matrix pattern. This pattern is transferred a bit at a time to a laser diode in the laser printer assembly. The laser creates an extremely tiny point of light. It is the laser that generates the dots that form the images. The laser beam is focused on a rotating mirror that causes the laser beam to be scanned across the drum. As the drum rotates, the bit pattern to be printed is applied to the laser, which turns off and on in response to the desired dot positions. Thus the page image is transferred to the drum one line at a time.

Another drum or roller near the drum in the printer is also electrically charged and attracts the fine granulated toner ink. The toner is a carbon-based powder containing iron and other materials that allow it to be attracted electrically and magnetically. The charged drum picks up the toner from the toner supply, giving it an identical electrical charge. Those areas of the drum that were exposed to light have a lesser negative charge than the more negative toner drum. In other words, the exposed area is positive with respect to the toner drum. Therefore, as the toner drum rotates near the main photosensitive drum, the toner is

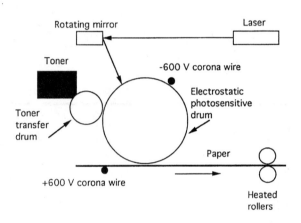

Fig. 14-10. Concept of a laser printer.

(continued next page)

attracted to those positive areas of the drum. This process transfers the toner to the drum in the desired image.

Next, a paper transport mechanism picks up one sheet of paper and passes it to the drum. As the paper moves through the transport mechanism, it passes by a corona wire that contains a high positive voltage. Thus the paper takes on a high positive charge. When the paper contacts the image drum, the toner is attracted and transferred to the paper.

Finally, the paper containing the desired image is passed through two heated rollers called the fuser. The very high temperature partially melts the toner and causes it to permanently adhere to the paper.

How is the image to be printed transferred to the drum?

    a. The drum is charged electrically.
    b. Toner is squirted at the drum to form the image.
    c. A pattern of dots is formed by turning the corona wire off and on at the right time.
    d. A laser is scanned across the drum as it rotates and is turned off and on to form a dot matrix pattern.

---

21   (*d.* A laser is scanned across the drum as it rotates and is turned off and on to form a dot matrix pattern.) Laser printers are the best printers available. They are fast and produce high-quality output. Low-cost laser printers are available that can print 6 to 8 pages per minute. High-speed laser printers are more expensive but can print at rates of 10 to 50 pages per minute.

Further, color laser printers are available. These use multiple toners of different colors and pass the paper through the printing process several times to combine the colors to create near-photographic quality print images.

Finally, most printers connect to personal computers by way of the popular Centronics parallel port. The information to be printed is transferred one byte at a time over the parallel port. Some printers have also been designed to use the RS-232 serial port on a computer, although this is less common. Newer printers will use the universal serial bus (USB).

In the past, most printers used which interface?

    a. RS-232
    b. USB
    c. Centronics
    d. IEEE 1394

---

22   (*c.* Centronics)   Go to Frame 23.

23 The other major category of peripheral devices for computers is mass storage devices. Mass storage refers to media capable of storing programs and data. The programs and data are permanently stored on such media and can be called into the computer when needed.

The RAM in the computer is simply not large enough to hold all of the programs and data with which the computer works. Typically when computers are used, only one program is present in the memory at any given time. Larger computers with multi-tasking operating systems permit two or more programs in memory simultaneously so that the computer can switch back and forth between programs as required.

The mass storage peripherals to be discussed in this section include the floppy disk drive, hard disk drive, and CD-ROM. Other, more specialized storage media, including the DVD drive, tape back-up, and PCMCIA mass storage units, are also covered.

There are several ways to categorize the mass storage devices used with personal computers. First, the storage medium may be either fixed or removable. A fixed storage media is one that is permanently in place, the most common example being the hard disk drive used in most personal computers. All other types of mass storage media are removable. These include the floppy disk or diskette, the CD-ROM and DVD disks, and the ZIP and JAZ drive disks. Magnetic tape cartridges, commonly used for back-up storage, are another example. Finally, there is the PCMCIA card that contains Flash ROM for limited mass storage. Most personal computers contain a floppy disk drive, a large capacity hard disk, and a CD-ROM drive. The other units are generally considered auxiliary add-on peripherals as required by the user.

Another way to categorize the peripheral devices is in terms of how the information is stored. The two major types of storage medium are magnetic and optical. Hard disks and all removable disks, such as the floppy disk and the ZIP and JAZ drives, are magnetic. Data is stored as tiny areas of magnetization on the surface of a disk coated with magnetic material.

The other form of storage is optical. Data is stored as pits in the surface of a disk; these pits are detected by a laser and photo optical detector. We will discuss all of these types in this section.

Which peripheral device is *not* usually a part of a standard off-the-shelf PC?

    a. CD-ROM
    b. floppy drive
    c. ZIP drive
    d. hard disk

Magnetic Recording and Playback Principles

25   In magnetic storage peripherals, the binary data to be stored is converted into electrical signals that are, in turn, used to create small magnetized areas on the surface of a disk or tape. The surface of the disk or tape is coated with magnetic material, a mixture of iron (ferrite) and cobalt that coats the entire surface of the disk.

To store a binary 1 or 0, an appropriate logic voltage is applied to a write or record head as shown in Figure 14-11. Voltage applied to the record head causes current to flow, producing a magnetic field in the core of the record head. A small gap is left in the core. As the magnetic disk or tape moves by the record head, the magnetic field produced in the core passes out of the core into the recording medium and back into the core through the gap. This process causes a small area of the magnetic medium to be magnetized. The area is somewhat similar to a very tiny bar magnet with north and south poles. The direction of magnetization is determined by the direction of current flow in the record head coil. If current is made to flow in one direction, a binary 1 is stored. If the current flows in the opposite direction, a binary 0 is stored.

The binary data recorded in this way is permanent. The surface of the magnetic media retains the data permanently unless it is deliberately erased by exposing the media to a very strong external magnetic field.

To access the data stored on the magnetic medium, a read head is used. Again, see Figure 14-11. Whenever the magnetic media passes through the gap in the read head, the small magnetic field is picked up and induces a voltage into the read head coil. The shape and polarity of the voltage produced by this determines whether the read head is detecting a binary 0 or binary 1. The small signal is amplified and changed back into standard logic voltage levels. In most cases today, a common head is used for both read and write operations.

Over the years as disk drives have become smaller and as recording densities have increased, a different type of read/write head has been developed. Called magnetoresistive heads, a constant current flows through these devices. When the head passes over a magnetized area on the storage medium, the resistance of the head changes. This causes the current in the head to change and a small voltage to be produced, indicating a binary 0 or a binary 1.

Magnetic storage media are generally impervious to external magnetic fields.

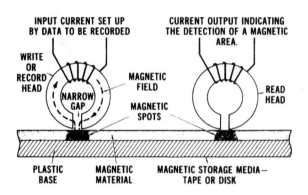

Fig. 14-11. Fundamentals of magnetic recording and playback.

a. True
b. False

Magnetic Disk Storage Format

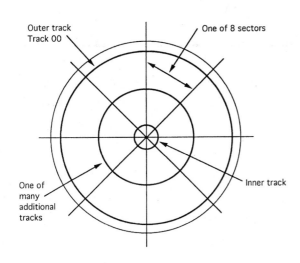

Fig. 14-12. **General concept of the data storage format on a magnetic disk.**

26 (*b.* False) Binary data is stored on magnetic disk drives as multiple concentric circles or tracks. See Figure 14-12. Each track is capable of storing many thousands of bits of data. To record and read back the data, a moveable read/write head is used. The head can be stepped or otherwise positioned over each track. Most disks have magnetic material on both sides, thereby doubling the amount of data that can be stored on an individual disk. A read/write head is provided for each side of the disk.

All data stored in a personal computer, whether it is in RAM, ROM, or a disk drive, is stored in byte-size units. One byte, or 8-bits, could represent an ASCII character or an 8-bit segment of a larger computer data or instruction word. Remember, most computer words are some multiple of eight bits, such as 16, 24, 32, and 64. Computer RAM is organized as areas for byte storage. The same is true on disk drives.

On a disk drive, data storage is also serial rather than parallel. The 8-bits of a byte are converted from parallel format into serial format usually by a shift register that produces parallel-to-serial data conversion. The serialized bits are then formatted in a special way (referred to as modulation in a disk drive) and then sent to the record heads.

In disk storage, the circular tracks are divided up into sectors with each sector capable of storing a fixed number of bytes. Again, see Figure 14-12. Floppy disks and early hard disk drives store 512 bytes per sector. Larger hard disks store 1,024 or 2,048 bytes per sector for increased storage capacity.

To simplify locating data on a disk, all disks are prerecorded (formatted) with binary codes that indicate the track or cylinder number, sector number, and number of bytes per track. Each track also contains spaces or gaps and special codes that help to divide up and identify the individual bytes of information. An error-correcting code, known as the cyclical redundancy check, is also recorded for each sector. This allows the disk drive controller to check for errors in recording or playback and, in some cases, to actually correct such errors.

How many bits of data can a disk with 200 tracks, 30 sectors per track, and 1,024 bytes per sector store?

a. 6,144,000 bits
b. 12,288,000 bits
c. 36,864,000 bits
d. 49,152,000 bits

27 (*d.* 49,152,000 bits) Simply multiply 200 tracks by 30 sectors and then multiply by 1,024 to get the total number of bytes that can be stored, or 6,144,000. To get bits, multiply by 8, or 49,152,000.

Go to Frame 27.

Floppy Disk Drive

28 The floppy disk drive (FDD) was the first magnetic mass-storage peripheral widely used with personal computers. The FDD was first invented back in the 1960s, but was never widely used until the 1970s. Today it is widely used in personal computers.

The floppy disk drive uses a removable media called a *floppy disk* or *diskette.* The storage media is a circular disk made of plastic coated with magnetic storage material on both sides. The disk is housed in a square cardboard or plastic cover. The cover provides a convenient way to protect the disk and to make it portable and removable. The disk drive itself is the mechanism into which the disk is inserted for reading or writing.

Diskettes are designated by their size and storage capacity. Early floppy disks were 8 inches in diameter. The first diskettes used with personal computers were 5-1/4 inches in diameter. Today the 3-1/2-inch diameter diskette is the only one used. The 8- and 5-1/4-inch floppy disks were packaged in a flat flexible cardboard housing (thus the term *floppy*) while today's 3-1/2-inch diskette is contained in a hard plastic housing.

Table 14-1 shows the size, format, and storage capacity of all of the most popular 5-1/4- and 3-1/2-inch disk drives used in personal computers. In all cases indicated, the disks store data on both sides. The number of tracks and sectors varies, but in all cases there are 512 bytes per sector. The 5-1/4-inch disks have disappeared from all modern PCs. The 1.44 MB 3-1/2-inch diskette is the most commonly used. The 2.88 MB 3-1/2-inch diskette is available but rarely seen.

Table 14-1. Specifications of Popular 5-1/4- and 3-1/2-Inch Disk Drives

Disk size	Total capacity (bytes)	Tracks	Sectors
5.25	368,640 (360 KB)	40	9
5.25	1,228,800 (1,200 KB or 1.2 MB)	80	15
3.5	737,280 (720 KB)	40	18
3.5	1,474,560 (1.44 MB)*	80	18
3.5	2,949,120 (2.88 MB)	80	36

*Most common

To determine the maximum storage capacity of a diskette, simply multiply the number of tracks by the number of sectors and by 512 bytes per sector. Then double that figure because the disk has two sides. For example, the most commonly used disk drive in personal computers today is the 1.44 MB capacity diskette. It has 80 tracks and 18 sectors with 512 bytes per sector. This gives a total of $80 \times 18 \times 512 \times 2 = 1,474,560$ bytes. Remembering that there are 1,024 bytes per kilobyte, this translates to $1,474,560/1,024 = 1,440$ KB. You can convert this to megabytes by dividing by 1,000 to give you 1.44 MB.

The most common diskette used in modern PC can store a total of

    a. 1.2 MB
    b. 1.4 MB
    c. 1.44 MB
    d. 2.88 MB

29

(*c.* 1.44 MB)   Two other important specifications of floppy disk drives are the *access time* and the *serial data rate.* The access time refers to the amount of time it takes for the floppy disk to find a particular piece of data on the disk. This is a function of how fast the read/write heads can move and the rotational speed of the disk. It takes anywhere from ten to several hundred milliseconds for the head in a disk drive to be positioned over the desired track. Data is then transferred serially from the track. The speed of the data is determined by the storage density, or how many bits per inch are contained on the disk. Rotational speed also determines data rate. For a 3-1/2-inch diskette, the rotational speed is 300 rpm. Some of the older 5-1/4-inch disks used 360 rpm. For standard high-density disks, the data transfer rate is 500 kbps or 5 million bits per second. Ignoring all of the blank spaces, addresses, and other special codes in the disk formatting and looking only at the data bytes themselves, the byte transfer rate is typically 45 KB per second.

The electronic circuitry associated with a floppy disk drive is generally divided into two major sections, the *drive electronics* and the *controller.* The drive electronics consists of the motor and its drive circuitry, the head positioning control circuitry, the read and write amplifiers, and some general control logic. All of this circuitry is mounted on a single printed circuit board packaged with the drive.

The controller is essentially the interface that connects the disk drive to the main computer bus. It consists of those circuits used to format and read all of the special codes on the disk as well as to perform serial-to-parallel and parallel-to-serial data conversions. All disk drives use the direct memory access (DMA) method of I/O.

In older disk drives, the drive electronics were packaged with the disk mechanism while the controller circuitry was contained on a separate printed circuit board that was plugged into a computer bus socket. The two units were

(continued next page)

connected by a ribbon cable. Today, the drive and controller electronics are usually packaged together inside the drive itself. A ribbon cable connects the drive and its internal controller to the computer bus.

The data to and from a disk is

a. serial
b. parallel

Hard Disk Drive

30 (*a.* serial)   The hard disk drive (HDD) or fixed disk is the primary data storage unit in a personal computer. It is used primarily to store the operating system and the applications programs used by the computer. However, it is also used to store data as the application requires.

Early hard disk drives appeared in the early 1980s and were capable of storing 5 to 10 megabytes of data. Today the typical hard disk drive in a personal computer can store many gigabytes (*giga* meaning billion).

The high storage capacity of the hard disk drive is made possible by the use of multiple storage disks as well as higher recording densities. Instead of using a flexible plastic disk, most hard disks are made of aluminum or ceramic and are not flexible. Two or more disks are mounted on a common spindle and permanently connected to the drive motor. Typical drive speed is 3,600 rpm, although new, smaller disks have rotational speeds up to 10,000 rpm. The entire unit is sealed and cannot be physically accessed.

The storage format of a hard disk is like that on a floppy disk, with data stored as concentric circles or tracks. When speaking of hard disk drives, the term *cylinder* is normally used to refer to the tracks. When multiple disks are used, each disk has two corresponding tracks, one on the top and one on the bottom. All of the same tracks on each disk in the drive taken together are collectively called a cylinder.

For example, assume that a hard disk drive contains four disks. Each of the four disks has two tracks, one on the top and one on the bottom. Track 0 (zero) is near the outer edge of the disk. Track 0 on the upper side of the disk lies exactly over track 0 on the bottom side of the disk. Directly below the top disk is the second disk with its two track 0s on the top and bottom. The remaining two disks also have a track 0 on the top and bottom, for a total number of eight tracks. These eight tracks form the 0 cylinder.

Most hard disk drives have hundreds of tracks or cylinders for storing data. These tracks or cylinders are further divided into a number of sectors. There are commonly 17 sectors per cylinder, although higher-capacity hard disk drives have more sectors per cylinder. There are commonly 512 bytes per sector, although the higher-capacity disk drives use 1,024 and 2,048 bytes per sector.

Computing the number of storage bytes on a hard drive is like that for floppy disks. First you multiply the number of cylinders by the number of sectors, and then multiply by the number of bytes per sector. Then you multiply that number by the number of read/write heads on the disks. For example, assume that a disk drive has 940 cylinders, 17 sectors and 512 bytes per sector. If you assume four disks with two heads each for a total of 8, the total storage capacity then is: $940 \times 17 \times 512 \times 8 = 65,454,080$ bytes. Generally this would be referred to as a 65 MB hard drive.

Most modern hard drives can store

a. kilobytes
b. megabytes
c. gigabytes
d. terabytes

---

31 (c. gigabytes) The access time for data in a hard drive also depends upon the speed of movement of the heads as they are positioned over the desired track and the rotational speed of the disk. Because access time is not fixed because of the rotational speed, only an average access time is given. Typical disk access times are from several milliseconds to 20 milliseconds, depending upon the disk drive. You will often hear the term *latency* used to express the access time to data on a disk. The speed of serial data transfer is also higher because of the greater storage density and the higher rotational speed. Data rates of several million bits per second are common.

The circuitry associated with a hard disk is generally the same as with a floppy disk drive. First, the drive electronics are associated with turning the motor and positioning the heads. The read/write amplifiers associated with the heads complete the drive electronics. The remainder of the circuitry consists of the interface or controller. The controller performs the serial-to-parallel and parallel-to-serial data conversions and all of the various formatting associated with labeling the cylinders and sectors.

In older drives, the drive electronics were mounted with the disk drive itself while the controller, or interface, was contained on a separate printed circuit board that was plugged into the computer bus. A ribbon cable interconnected the two. Today most hard disk drives contain both the drive electronics and the interface controller circuitry within the drive itself and it is attached to the computer bus by a ribbon cable.

Over the years, four different basic interfaces have been developed for hard disk drives. These are referred to as the ST-506, the enhanced standard device interface (ESDI), the integrated drive electronics (IDE), and the small computer systems interface (SCSI). The ST-506 and ESDI interfaces are no longer widely used. Most hard drives are of the IDE

(continued next page)

type where all the circuitry is packaged within the drive itself. SCSI interface drives are popular for very high density storage disks used in servers and other larger personal computers and work stations. In all cases, the interface uses DMA I/O.

Which hard drive interface is the most widely used?

    a. IDE
    b. SCSI
    c. ST-506
    d. ESDI

---

32   (*a.* IDE)   Go to Frame 33.

CD-ROM Drives

---

33   Another widely used mass storage peripheral in personal computers is the CD-ROM drive. The term *CD-ROM* stands for compact disk-read only memory. A CD-ROM is a plastic disk 4.72 inches in diameter and .05 inch thick that is identical in appearance to popular music CDs. However, the storage medium has been formatted for digital data rather than music. Programs and data are permanently stored on the CD when it is written, thus the designation ROM.

While most CDs are ROM, you can buy special CDs that can be written to with a special drive. These writeable CD drives normally use a laser to burn data into the CD surface. This is a one-time process in that the storage is permanent.

The great value of a CD-ROM is its massive storage capacity. Each CD can store up to approximately 650 MB depending upon how it is formatted. As software has gotten larger over the years, software publishers have switched to CDs as a way of initially distributing their programs to customers. Previously, programs were distributed on floppy disks, but the large programs today would require dozens of disks. So it is much less expensive to distribute software on a single CD.

Because of their large storage capacity, CDs can also store digitized audio, video, and graphics, as well as standard programs and data. This makes CD-ROMs the ideal distribution media for computer games. CDs are ideal for distributing encyclopedias containing text, color pictures, audio, and even short video segments. Today, virtually all personal computers contain a CD-ROM drive.

A standard CD-ROM diskette can be written on by the PC.

    a. True
    b. False

34 (*b.* False)   Data is stored and retrieved from a CD-ROM by optical techniques. The data to be stored is converted and encoded into a special serial format. The resulting digital data is then burned into the surface of the disk with a laser, permanently storing the data. A cross-section of the disk is shown in Figure 14-13. A clear polycarbonate plastic disk is covered with a thin metal layer and then a lacquer coating. The pit burned by the laser represents binary 0 while the unburned surface area, called the land, and represents binary 1. The information is recorded serially in a single, continuous spiral track from the inside of the disk to the outside.

To read the data stored on the disk, the light from a laser is shined on the surface of the CD as it is rotated. See Figure 14-13. The laser beam is reflected from the lands, sensed by a photodetector, and converted back into a binary 1 signal. The pits scatter the laser light so that little is reflected back to the photodetector. The resulting binary signal is decoded and then converted back into the original data. The laser, photodetectors along with the lenses, mirrors, and prism are packaged together in a housing that moves to follow the spiral data track as the disk rotates.

In most disk drives, the disk spins at a constant rate, 300 rpm in floppy disks and 3,600 rpm in disk drives. This method ensures constant angular velocity (CAV) when data is retrieved. With the spiral data track arrangement on a CD, initially it was desirable to produce a constant linear velocity (CLV). This meant that as the read head moved over the disk, the speed of rotation had to be changed to maintain a constant linear velocity. When the read head was positioned near the inner edge of the disk, it spun at

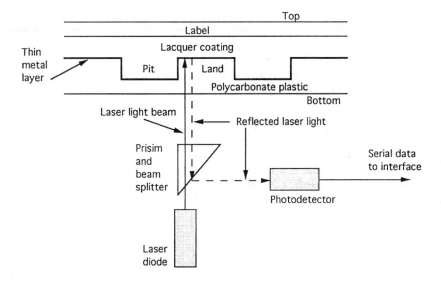

Fig. 14-13. Cross-sectional view of a compact disk showing how it works.

(continued next page)

approximately 500 rpm. As the read head moved to the outer part of the spiral near the outer edge of the disk, the speed slowed to approximately 200 rpm. This allowed a constant data rate to be retrieved. The serial data transfer rate of the original standard CD-ROM drive is 150 kBps.

Most CDs were found to be too slow in accessing programs and data. Manufacturers quickly responded by increasing the rotational speed of the disk. Double speed drives boosted the data transfer rate to 300 kBps. Drives that were four times (4X) faster with a transfer rate of 600 kBps were developed. This set off a trend of developing even faster drives with 10X, 12X, 20X, and 24X speeds. As this is written, 32X and even 64X speed drives are available. This increases the serial data transfer rate to 4.8 MBps. At these very high speeds, disk drive manufacturers switched from a CLV to a CAV speed format. Most high-speed drives have a constant rate of rotation regardless of the position of the read head.

The interface from the CD-ROM to the computer is usually by way of a standard disk drive IDE or SCSI interface. The interface plugs into the ISA bus on the PC motherboard.

Data is stored on a CD in

a. concentric tracks
b. a single spiral
c. multiple radius lines
d. vertical grooves

---

35   (*b.* a single spiral)   Go to Frame 36.

DVD Drives

---

36   DVD, meaning digital versatile disk, is an updated and improved CD-ROM. The storage media is a 120 mm (4.72 inches) optical disk like those used in standard CD-ROM drives. However, the storage and recording format has been changed to increase the storage capacity to 4.7 gigabytes. Additional versions of the DVD disk have been planned to allow storage of up to 17 GB.

The primary application of DVD disks is to record and play back video and music. A standard CD can only store approximately 74 minutes of video, which is insufficient for most movies. The new DVD disk contains far more space necessary to store a movie and related sound.

Some publishers of the media games and other entertainment programs found the storage capacity of a standard CD restricting. The new DVD storage format is designed to handle compressed video as well as Dolby AC-3 surround sound, which supports five channels of audio.

Although most DVD drives have been developed to connect to TV sets for playing both standard and surround sound movies, DVD drives are also available for personal computers. These units read standard format CD-ROMs as well as the higher-density DVD disks.

DVD disk drives can read CD-ROMs.

    a. True
    b. False

---

37   (*a.* True)   Go to Frame 38.

Backup Storage

---

38   It is often necessary to store large quantities of information for later access. Although the floppy disk is normally used for this purpose, its limited 1.44 MB capacity makes it inconvenient to store very large volumes of data. With standard PC hard drive capacities approaching 20 GB, enormous quantities of information can be stored. It is sometimes necessary to store segments of this information for transfer to another computer or as backup.

For example, suppose that it was necessary to store an 80 MB file of data. This would take approximately 56 standard floppy diskettes. This is expensive, inconvenient, and extremely time-consuming.

To meet such needs, several high-density portable hard disk drives have been created. These include the popular Iomega ZIP drive and the JAZ drives. A ZIP drive uses a removable 3.5-inch hard disk storage media that is the same diameter as a standard floppy diskette, but the package is thicker. Each disk is capable of storing up to 100 MB of data. A ZIP drive can be installed in any personal computer and used like a floppy disk or CD. A JAZ drive is also a removable hard disk but with a storage capacity of 1 GB.

Whenever it is necessary to store even larger quantities of information, a tape drive is the most appropriate. In many critical business and industry applications, it is desirable to back up the hard drive in case of damage. It is not uncommon for a hard drive to crash, thereby making it impossible to access the programs and data stored on it. To prevent just such a disaster, hard drives containing critical databases and other information should be backed up regularly. Backing up means copying the entire contents of the hard drive to another safer storage medium.

The medium most commonly used for this is a magnetic tape cartridge. These look very much like standard Phillips audio cassettes still widely used in consumer applications. However, these tapes store massive quantities of digital data, up to 12 GB uncompressed or up to 24 GB if data

(continued next page)

compression is used. Should a hard drive crash, the information can easily be restored by installing a new hard drive and transferring data from the tape back to the disk.

Storing the full content of a PC hard drive on another disk or tape is called

    a. redundant nonsense .
    b. data compression
    c. backing up
    d. CYA

---

39   (*c.* backing up)   Go to Frame 40.

Communications Interfaces

---

40 Communications interfaces refer to the circuits that a PC uses to talk to other PC by way of the telephone system or by interconnecting cables in local area networks (LANs). The two most common types communications interfaces are modems and network interface cards.

Modems

The term *modem* is short for modulator–demodulator. A modem is a special interface circuit that allows computers to communicate with one another and with larger distant computers and networks over standard telephone lines. Telephone lines, of course, were designed to carry analog voice information. To transmit digital data between computers over the telephone system, the digital data must be converted into an analog format. The modem must also be able to convert the received analog signal back into the original digital data.

Refer to Figure 14-14. The modem consists of two parts: the transmitter (TX) or modulator, and the receiver (RX) or demodulator. The modem is usually mounted on a single printed circuit board that plugs into the bus of the PC. Another less-used type of modem is an external modem that connects to one of the existing I/O ports on the PC (i.e., RS-232). The modem connects to the telephone line by way of a twisted-pair cable and the common RJ-11 telephone connector. The telephone system carries the data to the remote computer that also has a modem. The transmitting and receiving modems must be set so that their speeds and modulation methods match.

The modulator part of a modem converts the digital data into specially formatted audio tones that pass easily over the telephone network. At the receiving computer, the modem picks up the special analog coded information and translates it back into the digital data. This is referred to as *demodulation.*

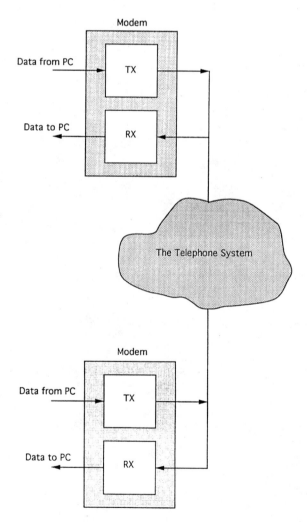

Fig. 14-14. Data communications by modem.

The very earliest modems used with computers back in the 1970s used a form of frequency modulation known as frequency shift keying (FSK). The binary data was transmitted by sending one frequency audio tone for binary 0 and another tone for binary 1. This method allowed data to be transmitted at a rate of 300 bits per second (bps). Later modems using phase-shift keying (PSK) permitted higher data speeds of 600, 1,200, and 2,400 bps data rates.

The greatest limitation to transmitting digital data over the telephone system is the speed of transmission. The speed of transmission is limited by the bandwidth of the telephone system, which is typically less than 4kHz. As computers have become more powerful and as programs and data have become larger, it takes longer and longer to transmit this information over the telephone system. Although short files can be transmitted almost instantaneously, longer files take many seconds or even many minutes to transfer. The transfer rate has become unacceptable in many modern applications. The wide adoption of the Internet as a method of accessing information has put increasing pressure on modem manufacturers to increase the data transfer speed.

Continuing developments in modems rapidly increased data transmission rates to 9,600 bps and then later to 14.4 kbps, 28.8 kbps, 33.6 kbps, and 56 kbps. These high-speed data formats use a special sophisticated form of quadrature amplitude modulation (QAM), which is a combination of amplitude modulation (AM) and phase-shift keying (PSK). Some forms of modulation also use a data compression technique that also increases speed.

All of the modulation techniques are defined and standardized by the International Telecommunications Union (ITU), a consortium of countries that establishes standards that make data communications over the telephone system compatible. The ITU standards are designated by a letter number combination as shown in Table 14-2. The V.xx designations are the standard numbers that define the modulation and transmission methods. The related speed of data transmission is given.

Table 14-2. International Modem Standards

Standard	Modulation	Speed	Other
V.22	PSK	1200 bps	
V.22bis	QAM	2400 bps	
V.32	QAM	9600 bps	
V.32bis	QAM	14.4 kbps	Trellis coding*
V.32terbo	QAM	19.2 kbps	Trellis coding*
V.34	QAM	28.8 kbps	Trellis coding*
V.42bis	QAM	33.6 kbps	Data compression, error detection
V.90	QAM	56 kbps	Combined standard for 56 kbps

*Trellis coding is a special form of QAM that increases speed of transmission.

(continued next page)

More recently, a 56 kbps modulation method was developed. However, two competing standards were created. These were incompatible with one another. Recently the ITU combined the two and established a single compatible standard designated V.90.

Modems with 56 kbps transmission rates represent the maximum data speed that can be obtained by analog modulation methods over the telephone system. Even then those speeds are rarely achieved in practice. If a modem cannot connect to another modem at its maximum speed, it automatically backs off and drops to lower and lower speeds as it adapts to the line conditions and any noise that is present. Although a modem may be capable of communicating at 56 kbps, most modems will often connect at a lower speed, such as 33.6 or 24 kbps.

The most common form of modulation used with conventional analog modems is

a. FSK
b. PSK
c. ASK
d. AM plus PSK

---

41 (*d.* AM plus PSK)   QAM is a combination of AM plus PSK.

Higher data transmission speeds can be achieved by using an ISDN line and interface. ISDN stands for *integrated services digital network.* This is a system created by the telephone companies for data transmission. Most telephone companies can now supply a separate ISDN line to homes and businesses for data communication. This is an all-digital line, meaning that information is transmitted by digital pulses rather than by analog modulation methods. Data speeds up to 128 kbps can be achieved on an ISDN line. A special ISDN interface or modem is required for connection. Currently ISDN is not very popular because it is not available everywhere and it is expensive.

In the quest for even higher speeds, many new data communications techniques are being developed. Two of the most promising are cable modems and the ADSL modem. The cable modem uses the cable TV system of coax and fiber optic cable to transmit digital data. The digital data is modulated onto a higher frequency carrier that is transmitted over the cable along with television signals. Very high data rates can be achieved. It is not uncommon for cable modems to transmit at speeds up to 10 Mbps.

The most promising new data communications method is referred to asymmetric digital subscriber line (ASDL). This new system uses standard analog telephone lines to transmit high-speed digital data. The term *asymmetric* refers to the fact that the speed of transmission into the computer (download) is significantly higher than the transmission from the computer to an external destination (upload). Incoming or downstream data can be as high as 8

Mbps, while outgoing or upstream can reach as high as 1 Mbps. There are other variations. Many believe that ADSL modems will be the new standard in PCs in coming years.

Finally, very high data rates can also be achieved with satellite systems. Many of the newer digital TV satellite services include a data transmission facility with data rates as high as 400 kbps.

Which modem type will most likely be the most popular in the future?

    a. ADSL
    b. ISDN
    c. Satellite
    d. cable TV

Network Interfaces

42   (*a.* ADSL)   Most personal computers have interfaces that enable them to communicate with other computers. A modem connection is the most common, and the major application is access to the Internet and World Wide Web. A LAN connection is the next most common connection. A LAN is a local area network, an interconnection of PCs in an office, building, department, or campus for the purpose of implementing e-mail, sharing peripheral devices such as fast laser printers, or accessing large databases. Usually a fast PC with a large amount of RAM and hard disk space is used as the controller or central communications computer for the network. This PC is referred to as the *server.* It runs a network operating system (NOS) software and other applications like e-mail and databases. The server and the NOS coordinate all LAN operations. It is estimated that over 70 percent of all PCs in organizations are networked.

What is the name given to the PC that controls the LAN?

    a. central PC
    b. server
    c. LAN manager
    d. control PC

43   (*b.* server)   All PCs are connected to the LAN by special interfaces and cables. Many different interfaces and methods of cabling have been developed over the years, but the most common today is called Ethernet. The original Ethernet used coax cable and ran at 10 Mbps and was referred to as a 10BASE-2 or 10 BASE-5. Today, most Ethernet LANs use twisted-pair cable and run at 100 Mbps and are called 100BASE-T. A 1 Gbps Ethernet is being developed.

(continued next page)

The topology or physical connection of an Ethernet LAN is a bus. All of the PCs share a common coax or twisted-pair bus. See Figure 14-15. That means that only one PC can transmit at a time to avoid collisions. Ethernet uses an arbitration system called Carrier Sense Multiple Access/Collision Detection (CSMA/CD). If two PCs try to send at the same time, the first to transmit wins the bus and the other backs off and waits a short time before transmitting again. Data is transmitted in bursts or packets of data up to 1,500 bytes long.

Each PC contains an Ethernet network interface card (NIC). This card plugs into the PC bus and handles all serial data reception and transmission. Some interfaces are built into the motherboard. NOS software is installed on each PC so that it can talk to the server. The most commonly used NOS are Novell's NetWare and Microsoft's NT. In general, the PC user does not use the LAN while he or she is doing work with spreadsheets, word processing, and so forth. The NIC and NOS are transparent to the user. They come into play only if the user deliberately tries to access a shared printer connected to the server, a database installed in the server hard drive, or the e-mail system. Some servers contain modems that connect the LAN to the Internet, thus eliminating the need for each PC to have a modem.

What is the original speed of the Ethernet LAN?

    a. 4 Mbps
    b. 10 Mbps
    c. 100 Mbps
    d. 1 Gbps

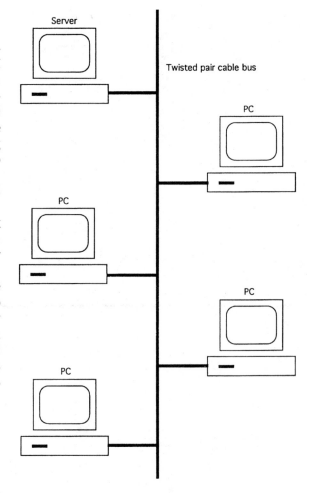

Fig. 14-15. An Ethernet bus LAN.

44   (*b.* 10 Mbps)

Other Peripheral Devices

The peripherals covered in this chapter are the most common, but there are many others. Examples of other peripheral devices include the following:

- Sound: An interface that plugs into a PC bus and allows the PC to play stereo audio on small speakers from CD games or regular audio CDs.
- Scanner: A peripheral that converts text, graphics, or photos to a digital file that can be incorporated into another document.
- Bar Code Reader: A device used to scan standard bar codes into the PC.
- Video Capture: An interface containing a fast analog-to-digital converter that will digitize video from a TV set, VCR, or video camera.
- Joy Stick: A manual control used with many PC games.
- Plotter: A device that draws large graphics with ink pens on paper or clear vinyl plastic.

Index